# About Island Press

Since 1984, the nonprofit organization Island Press has been stimulating, shaping, and communicating ideas that are essential for solving environmental problems worldwide. With more than 1,000 titles in print and some 30 new releases each year, we are the nation's leading publisher on environmental issues. We identify innovative thinkers and emerging trends in the environmental field. We work with world-renowned experts and authors to develop cross-disciplinary solutions to environmental challenges.

Island Press designs and executes educational campaigns, in conjunction with our authors, to communicate their critical messages in print, in person, and online using the latest technologies, innovative programs, and the media. Our goal is to reach targeted audiences—scientists, policy makers, environmental advocates, urban planners, the media, and concerned citizens—with information that can be used to create the framework for long-term ecological health and human well-being.

Island Press gratefully acknowledges major support from The Bobolink Foundation, The Curtis and Edith Munson Foundation, The Forrest C. and Frances H. Lattner Foundation, The Freedom Together Foundation, The Kresge Foundation, The Summit Charitable Foundation, Inc., and many other generous organizations and individuals.

The opinions expressed in this book are those of the author(s) and do not necessarily reflect the views of our supporters.

# Squirrel

# Squirrel
## How a Backyard Forager Shapes Our World

Nancy F. Castaldo

ISLANDPRESS | Washington | Covelo

© 2025 Nancy F. Castaldo

All rights reserved under International and Pan-American Copyright Conventions. No part of this book may be reproduced in any form or by any means without permission in writing from the publisher: Island Press, 2000 M Street, NW, Suite 480-B, Washington, DC 20036-3319.

Library of Congress Control Number: 2025934652

All Island Press books are printed on environmentally responsible materials.

10 9 8 7 6 5 4 3 2 1

*Keywords:* squirrels, environment, nature, biodiversity, forests, trees, rodents, wildlife, animals, history, keystone species, forest ecology, urban wildlife, climate change, squirrel biology, living with squirrels, why do squirrels collect nuts, how to keep squirrels from my bird feeder

*For my friends,
my "scurry" of support, laughter, and joy.
Thank you for always being the nutty, wonderful, and endlessly
inspiring force that makes life anything but common.*

"So far as our noblest hardwood forests are concerned, the animals, especially squirrels and jays, are our greatest and almost only benefactors. It is to them that we owe this gift. It is not in vain that a squirrel lives in almost every forest tree or hollow log or wall or heap of stones."
—Henry David Thoreau

"The last word in ignorance is the man who says of an animal or plant: 'What good is it?'"
—Aldo Leopold

"For it is the right and property of all natural objects, of all genuine talents, of all native properties whatsoever, to be for their moment the top of the world. A squirrel leaping from bough to bough, and making the wood but one wide tree for his pleasure, fills the eye not less than a lion—is beautiful, self-sufficing, and stands then and there for nature."
—Ralph Waldo Emerson

## Contents

| | |
|---|---|
| *Author's Note* | *xiii* |
| *Preface* | *xv* |
| | |
| Chapter 1: As a Keystone Species | 1 |
| Chapter 2: In the World | 16 |
| Chapter 3: As a Cultural Icon | 32 |
| Chapter 4: As a Mammal of Mystery | 41 |
| Chapter 5: As an Ambassador | 51 |
| Chapter 6: As a People Greeter | 68 |
| Chapter 7: As an Interloper | 75 |
| Chapter 8: On the Continent | 97 |
| Chapter 9: As Dinner | 108 |
| Chapter 10: As a Marauder | 114 |
| Chapter 11: On the Move | 127 |
| Chapter 12: By the Numbers | 138 |
| Chapter 13: As a Shape-Shifter | 144 |
| Chapter 14: As an Endangered Species | 154 |
| Chapter 15: In the Future | 171 |
| | |
| *Acknowledgments* | *183* |
| *Coexisting with Squirrels* | *185* |
| *Squirrel Family Tree Classification* | *187* |
| *Source Notes* | *195* |
| *Bibliography* | *221* |
| *Illustration Credits* | *231* |
| *About the Author* | *233* |
| *Index* | *235* |

## Author's Note

In 1973, Frederick S. Barkalow Jr. and Monica Shorten wrote *The World of the Gray Squirrel*. For years, the book was one of the only nonfiction titles focused on the species. The two authors began the book with a note justifying their writing about the ubiquitous squirrel species. "Our purpose in writing this book is to present the squirrel as we see it: a lively versatile rodent, neither 'good' nor 'bad,' a creature about which there is still plenty to learn as it strives to adapt to the bewildering changes man brings to its world."

If asked why I decided to follow in their footsteps and tackle the subject of squirrels, I'd agree that there is still plenty to learn. In addition, a lot has changed for our climate and our global squirrels since 1973, when their book was released and I was a child hand-feeding chipmunks. Since then, I've studied and written about many charismatic, rare creatures, including endangered California condors and Galápagos tortoises. It is easy to spotlight species teetering on the brink of extinction, those poster wildlife species that scream for attention. However, there is so much more that we need to discover and learn about the wildlife species that are in the background, such as our seemingly ubiquitous, mysterious squirrels. It is time to bring these rodents out of the white noise of our world and focus on them while we are fortunate to still have many of their species all around us.

This book is an attempt to focus some attention on the squirrels that fill our lives with joy and frustration. I've chosen to share some of my experiences in coming to grips with how I perceive squirrels in my own life and how others I meet perceive them in theirs. I've selected stories and places that illustrate the nature of squirrels in our environment and in our lives with the clear intention of inspiring my readers to consider our global squirrel species in this uncertain world we all inhabit.

Published by W. Darton, J. Harvey, & W. Belch. London. July 1.1799.

COMMON SQUIRREL.

## *Preface*

The world is waking. The morning sun shines a pink light across the hill to my right as I step out my back door with my dog. Birdsong is minimal today, with spring hanging out in the near distance and winter waning. Warm days and cold nights wake the sap in the maples, including the old one to my left that casts its shadow on my roof. As we take it all in, a scamper of red squirrel fur alights on the maple branches to my left and flies from the tree, across the grass, to the trees edging the woods. From the safety of those trees, it lands on a branch and begins chittering at us for interrupting its morning.

It's early spring, and the little American red squirrel is busy. My heart is glad.

I'm not alone in my observation and emotion. Since antiquity, squirrels have fascinated and frustrated humans, entertained and exasperated. After all, squirrels, from the Sciuridae family, appear everywhere. They are in our backyards and parks, on our streets, at our feeders, and in our attics. They amuse us as they chase each other like kids playing tag. They scurry up and down trees. They chit and chat as if scolding us as we walk through woodlands. For many of us, not a day goes by without seeing one. Most children can identify a squirrel because squirrels are often the first wild mammal we encounter in our lives, and their images fill childhood storybooks.

Perhaps you, like actor and comedian Ellie Kemper, who shares her childhood obsession with squirrels in her memoir, *My Squirrel Days*, have your own childhood squirrel story to share. Whether you were a keen observer of local squirrel populations, or fed a squirrel in a local city park, or even chased one from a bird feeder, you likely have a squirrel tale buried in your childhood memories.

The fact that squirrel species span the globe means that a child in America is as likely to see a squirrel as a child in a European park, sometimes even the same species, as is the case with North American gray squirrels. According to the IUCN (International Union for

Conservation of Nature and Natural Resources) Red List of Threatened Species, there are currently 294 identified squirrel species, including subspecies and varieties. Indigenous to the Americas, Eurasia, and Africa and introduced into Australia, species include arboreal (tree), ground, and flying squirrels. Active during the day, diurnal squirrels can be observed in both bustling urban parks and the most remote rural landscapes.

Because squirrels are found everywhere, they are often passed over in formal research for more exotic or endangered wildlife species. Their prevalence and close association with humans give them a kind of anonymity even as they suffer significant mortality from cars, pesticides, and other human-related causes. This happens with such frequency that it can feel as if squirrels are disposable, like insects on a windshield. But roadkill and habitat loss significantly impact their survival as a species, and we know that wildlife is never disposable.

So many threatened Indian giant squirrels are killed by speeding vehicles in India's forest regions that the government has stepped up measures to protect them, such as constructing wildlife crossings. Human-caused impacts to habitats, some from climate change and others from habitat loss, affect every squirrel species. The challenges are numerous.

So, given our close and frequent association, is anyone in areas other than India considering squirrels? Even though many squirrel species have lived alongside us for millennia, they're curiously understudied and underappreciated. I've been thinking about this a lot, and it is my reason for writing this book.

"There's actually so little known about them ecologically," said Noah Perlut, professor and director of Project Squirrel at the School of Marine and Environmental Programs, University of New England, to *The Washington Post*'s Kate Morgan in February 2024. In addition, Perlut's Project Squirrel website claims, "While unquestionably the most commonly seen wild mammal in New England, the gray squirrel is entirely unstudied." Perlut doubled down on this when I later met up with him on the Maine campus. "They've been studied pretty well in places where they have been introduced because they're causing ecological havoc, but in their native range there is remarkably little known about their behavior. When I say behavior, even their basic ecology."

Perlut is not the only scientist who thinks this. Thaddeus McRae, a biologist at Lee University in Tennessee, said it best: "Some people are bird people, some people are cat people. Some people love bugs. That can influence choices of what gets studied as much as anything else. Squirrels are cute, but so commonplace to many of us that they become background."

Some squirrels seem to beg to be studied. Diurnal tree squirrels, for example, are ideal subjects for scientists who explore the behavioral decisions involved in consuming and storing food and how squirrels influence plant reproduction and forest regeneration. They are also important subjects for investigating the evolution of social structure among mammals, reproductive strategies, and population biology.

Even so, many are understudied. There are, however, researchers working to uncover more about these essential rodents. The term *essential* might seem too strong to describe animals that don't appear to be keystone species. Most people would find it difficult to imagine an ecosystem collapsing, like a house of cards, without them. But in this book I attempt to demonstrate why squirrel species are essential to the fabric of our global ecosystems and worthy of a closer look. Many are keystone species in their environments, propagating woodlands through their caching behaviors and modifying habitat in ways that are useful to other species. Squirrels are also a source of food for non-human predators such as coyotes, snakes, hawks, and owls. "We don't really think about their more complex roles in our ecosystems and our daily lives," said Alex Potash, a postdoctoral researcher in the University of Florida's Department of Wildlife Ecology and Conservation.

As we enter an era of worldwide biodiversity loss, we should pay attention to common species such as squirrels as well as more iconic, charismatic species because they all have a limit to the stress they can absorb. It is vital to study common species before their populations begin to decline or crash outright. Who would have imagined a century ago that passenger pigeons, once numbering in the billions, would disappear, or that in the 1970s the common American alligator would face extinction?

Even as I write, two squirrel species are classified as endangered under the Endangered Species Act of 1973: the Carolina northern flying squirrel (*Glaucomys sabrinus coloratus*), identified by its bright cinnamon-brown dorsal fur, the gray fur around its face and at the end of its tail, and its bicolor belly fur; and the Mount Graham red squirrel (*Tamiasciurus hudsonicus grahamensis*), grayish brown with rusty and yellowish markings on its back, a white belly, but no white-fringed tail. Many more are in similar straits. Others can be found on state endangered species lists.

Still many more romp, scurry, and sploot across our woodlands and parks, making substantial contributions as members of their ecosystems. Our forests depend on squirrels for regeneration: Squirrels cache acorns and other nuts and seeds for food but overlook enough to allow for seedlings to sprout in new locations.

Well-known biologist John Koprowski, a professor in the Haub School of Environment and Natural Resources at the University of Wyoming, has studied squirrels for decades and agrees that "squirrels are excellent and understudied models for science and conservation . . . even the most common species are not all that well studied." Some might refer to Koprowski as a squirrelologist. He expounded on the importance of squirrels during a 2015 interview in Finland, where he gave the opening plenary for the Seventh International Colloquium on Arboreal Squirrels, saying that "humans can [plant seeds], but squirrels are exceptionally good at doing that. And they do it for free."

Caching seeds and nuts is a vital way squirrels influence the forest ecosystem. Geologist Ellen Wohl claims, in her book *Something Hidden in the Ranges: The Secret Life of Mountain Ecosystems*, that the caches of tree seeds, called middens, created by squirrels in Rocky Mountain National Park serve as a repository for those seeds should a wildfire burn the forest's living vegetation. The squirrels' middens—piles of stored cones and discarded scales—form the actual backbone of the forest ecosystem. While still more needs to be explored about squirrels' caching behavior, we also know that squirrels move around essential underground fungi that help tree growth and form a foundation for much of a forest. Squirrels spread truffle spores through their digestive systems onto the forest floor, allowing the fungi to create critical symbiotic relationships with the nearest tree roots. In addition, ground squirrels' digging pulls important fungi to the surface and can help with forest soil recovery.

Not only are squirrels crucial components of their forest ecosystems; they also—unbeknownst to them—serve in the human realm as our country's ambassadors and "public pets." The charismatic rodents, according to University of Pennsylvania science historian Etienne Benson, were often brought from nearby rural woodlands to cities to "create pockets of rural peace and calm" for early America's first urban settlers.

But in a world of climate upheaval and habitat loss, could squirrels, beloved and notorious, become as ill-fated as the common, yet extinguished, passenger pigeon? According to Rosalie Barrow Edge, conservationist and founder of Hawk Mountain Sanctuary in Pennsylvania, "The time to protect a species is while it is still common."

It is time to consider squirrel species, common and uncommon, in science, our lives, and our forest management so that they do not face the same fate as so many other species we've overlooked and lost, further putting our global environment in peril.

*Chapter One*

# As a Keystone Species

SCOIATTOLO DEL CANADA

> *"Keystone: a central stone at the summit of an arch, locking the whole together."*
> —Oxford English Dictionary

"I learned a fun fact about squirrels. . . . Squirrels cannot find 80 percent of the nuts they hide. Are you kidding me? Is that the greatest thing you've ever heard in your life? First of all, animals aren't supposed to make mistakes. But secondly, I made this realization. . . . Hold your skulls in, because your brains are gonna [f***ing] explode. That's how trees are planted," pronounced comedian Sarah Silverman.

Ernest Thompson Seton, naturalist and pioneering founder of the Boy Scouts of America, wrote in the foreword to his 1922 animal-fiction story *Bannertail: The Story of a Graysquirrel*, "In the nut forests of America, practically every tree was planted by the Graysquirrel, or its kin. No squirrels, no nut-trees." While Seton often anthropomorphized his animals with the human qualities of curiosity, desire, and sympathy, garnering negative and vocal criticism from other naturalists, his observations should not be discounted.

Is Seton or Silverman correct? Perhaps both or neither. Plants rely on seed dispersal for their survival. While some plants have seeds that fly away in the wind, many need a creature, such as a bird or squirrel, to carry away the seed. Whether a seed is picked up on fur and brought inadvertently to a new location, excreted after being eaten, or carried away and planted in the earth, plants have relationships with dispersers. These relationships are crucial to the survival of forest ecosystems; they are crucial to our own survival. Evidence of a seed dispersal crisis in Europe in 2024 indicated that for 30 percent of plant species, most of their dispersers are threatened or declining, demonstrating how vital seed dispersers, such as squirrels, are in our world.

Squirrels are dispersers. They do plant trees, but the process is more complex than both Silverman's and Seton's pronouncements. Squirrels have used their skill at finding food to their advantage, and this demonstrates an "enhanced" cognitive ability. Their food-caching behavior does not lack purpose or direction. And so we must consider the brains of squirrels, how squirrels contribute to the health of our forests, and their role as keystone species.

"Not much goes on in the mind of a squirrel. Huge portions of what is loosely termed 'the squirrel brain' are given over to one thought: food," wrote Kate DiCamillo in *Flora and Ulysses*, a novel centered on a girl and a squirrel that won the John Newbery Medal in 2014. DiCamillo wasn't wrong in describing squirrels' focus and behavior toward food. But that drive for sustenance is present among most animals. From wolves to pelicans, wild creatures constantly think about their next meal. It's a matter of survival. Squirrels aren't an exception. Without that drive for sustenance, they perish. But there is more to the squirrel brain than DiCamillo's description suggests.

*Squirrel brain* is often used as a derogatory metaphor for someone having scattered and fuzzy thinking without much depth or connection. The truth is much different. Squirrel brains, about the size of a walnut, are substantially more complex. Just as scientists have observed in other creatures, there is a great deal we are uncovering about animal brains and cognition. Squirrel brains are quite large compared with the size of their bodies, quite larger than those of other rodents. This ratio, comparable to that of many primates, demonstrates cognition complexity. This complexity is seen in the way they group their nuts. Squirrels use the method of "spatial chunking," keeping their nuts grouped according to type, to help them remember where they are stashed. Walnuts with walnuts. Hazelnuts with hazelnuts. They also group nuts by size. This is not fuzzy thinking.

Noted US squirrel researcher Michael Steele, along with researchers from Germany and the United Kingdom, explored the "enhanced" cognitive ability of gray squirrels. While they found that some cognitive abilities in gray squirrels, such as solving novel problems, has undergone mild variation as they have adapted to new environments, the previously reported enhanced performance is likely a general characteristic that brings fitness advantages to this species and contributes to their adaptability to new environments. Squirrels demonstrate exceptional problem-solving abilities and have a complex communication system that includes both sound and scent to share information of threats and food sources. "Squirrel brain," then, is far from scattered and fuzzy. And having that brain also has a huge impact on our forests.

In *Bannertail*, Seton captured a squirrel's robust autumn food drive:

"No longer wabbly or vague, as in that first autumn, but fully aroused and dominating was the instinct to gather and bury every precious, separate nut. Bannertail had had to learn slowly and partly by seeing the Redsquirrels making off with the prizes. But he had learned, and his brood had the immediate stimulus of seeing him and their mother at work; and because he was of unusual force, it drove him hard, with an urge that acted like a craze. He worked like mad, seizing, stripping, smelling, appraising, marking, weighing every nut he found."

Anyone, like Seton, who has observed squirrels during autumn can relate to Bannertail's urge that made him act in a "craze" to gather nuts. The urgency squirrels exhibit not only enables them to feed throughout the cold months ahead; their behavior also builds forests.

This wasn't the first time Seton wrote about squirrels. A decade earlier, he included squirrels among sixty mammalian wildlife species in *Life-Histories of Northern Animals: An Account of the Mammals of Manitoba*. Most of his remarks that follow the sections on classification are his personal observations or those of his friends and colleagues. "I am informed by A. K. Fisher that at the southern end at Lake George, in early autumn, it is sometimes an everyday occurrence to see Red-squirrels swimming across the lake, from west to east (about two miles)—never in the opposite direction. The chestnut grows abundantly on the eastern side of the lake, but it is comparatively scarce on the western, and these extensive migrations always take place in years when the yield of chestnuts is large."

Lake George is a thirty-two-mile-long lake in northern New York state's Adirondack Mountains. The mountains make up the southern part of the Eastern Temperate Forests ecoregion, which extends into Maine and eastern Canada and is home to the state's Adirondack Park Forest Preserve. The forests in the park and around Lake George consist of hemlocks, spruces, beeches, pines, and broad-leaved trees. Red squirrels, as well as eastern gray squirrels and two species of flying squirrels, make their home in the forests along with a host of other mammals, including moose, deer, and beavers.

A. K. Fisher's observations of chestnut-searching swimming squirrels in Lake George were not the only observations in the Adirondacks. Early studies of red squirrels in the mountains reported sightings of squirrels swimming across Big Moose Lake, Long Lake, Brantingham Lake, and Lake George. Seton also included James Higby's observation of June 1877; Higby witnessed as many as fifty squirrels crossing Big Moose Lake.

Another account was documented in Winslow Watson's history of Essex County: "The autumn of 1851 afforded one of these periodical invasions of Essex County. It is well authenticated, that the red-squirrel was constantly seen in the widest parts [about seven miles] of [Lake Champlain], far out from land, swimming towards the shore, as if familiar with the service; their heads above water, and their bushy tails erect and expanded, and apparently spread to the breeze. Reaching land, they stopped for a moment, and relieving their active and vigorous little bodies from the water, by an energetic shake or two, they bounded into the woods, as light and free as if they had made no extraordinary effort."

The massive chestnut trees at the Adirondack lake were appreciated not only by native squirrel species; they also appeared in many works of art at the turn of the century, including Robert Melvin Decker's *Old Chestnuts at Bolton, Lake George*, circa 1890–95. Alfred Stieglitz captured a dying chestnut tree on the eastern edge of Lake George in a 1927 photograph. None of this art captured swimming or climbing squirrels.

However, those chestnut trees, a boon for American red squirrels, were doomed when the nonnative Chinese chestnut tree entered the United States with a lethal fungus.

On an autumn day roughly a century after Stieglitz took that photo, I sat near that site at the southern end of Lake George, reading Diane Ackerman's passages about squirrels in her book *Cultivating Delight*. As with my many other visits to the lake over the years, I did not witness a single red or gray squirrel swimming in the lake. I also haven't seen those majestic chestnuts that propelled the swimming behavior that Seton memorialized in 1909. The lack of chestnut trees altered the composition of the forests and the behavior of the wildlife that depended on them.

Fortunately, those American red squirrels did not completely rely on the chestnuts for their survival. But it is easy to see how we could have had a different outcome with the population of our common squirrels.

These days, post chestnut tree habitation, red squirrels harvest seeds from Adirondack conifers, including pines, spruces, and firs. They also feed on birch catkins and sugar maple bark and seeds. The Adirondack forests also provide them with nuts from beaked hazelnut and American hazelnut and the berries of northern wild raisin, wintergreen, and partridgeberry. They are regularly seen on all forest trails. The relationship between the American red squirrels and the forests of New York's Adirondack Mountains was also explored in the *Roosevelt Wild Life Annals*. In 1929, field naturalist Robert Hatt wrote about this relationship and our relationship with squirrels in "The Red Squirrel":

> The opinions of people, are, however, anything but unanimous when the status of the squirrel in relation to the community at large is considered. Some see in this impetuous creature only a vivacious forest sprite whose sole claim to existence lies in its charming disregard for all ordinary customs and civilities of life. Others see the squirrel as a destroyer of vast quantities of buds and seeds, a creature barking trees, robbing birds' nests, and driving the more desired gray squirrel from its territory; an animal of little interest as game; a rodent perhaps best dispensed with. Others again note the squirrel burying seeds and nuts for use in the season of scarcity and conclude that, though all unconsciously, it is actually aiding in natural reforestation. Somewhere among these various and conflicting views lies the truth. Like all other wild creatures, the red squirrel has a myriad relations with its environment, some of which are of benefit to a man's community as a whole, some suit the interest of but a few, and others are in themselves wholly undesirable. For this reason the red squirrel is alternately denounced and defended.

In their natural habitat, native squirrel species are a boon for woodlands. That strong relationship can be observed every autumn as they answer the urge to gather food.

While Europe's native Eurasian red squirrels have been front and center in their competition with their North American cousins, the squirrels there and everywhere have all been working steadily at fulfilling their ecological niche by consuming and dispersing seeds. The two red squirrel cousins are similar in that they are smaller than their gray squirrel relatives. Although they are smaller, the reds are equally important in building forests.

Hatt's missive on the importance of squirrels mirrors the message evoked in the ancient Sanskrit tale the *Ramayana*: No matter how small a creature is, it can make a difference. Squirrels of all species, small and ever present, do much to strengthen the health of our forests.

Solely examining our oak forests demonstrates their importance. A gray squirrel's acorn-caching behavior for food not only benefits the squirrels themselves; since many of those seeds are never consumed by the squirrels, the acorns become the source of new oak trees in the forest. A single oak tree functions as a lifeline to countless other creatures in the forest, including birds, raccoons, snakes, opossums, lizards, moths, butterflies, wasps, and so many more species that depend on leaf litter for shelter and nourishment, demonstrating how squirrels can serve as a keystone species in their forest ecosystem.

By using a scatter hoarding method of burying single nuts in lots of locations, squirrels provide one of the only ways nuts get far enough from the parent tree's shade to have a successful germination. The eventual oak germination leads to the growth of trees with extensive underground root systems that far exceed the width of the tree's canopy. These underground root systems not only stabilize soils but also contribute to carbon sequestration and watershed management. Therefore, squirrels, regardless of species, contribute to forest health in various ways.

While all squirrel species are useful, some are more successful for certain tree species. Take, for instance, New Mexico animal ecologist Jake Goheen's prediction that seven times as many walnuts germinate when scatter hoarding gray squirrels gather them compared with those hoarded by red squirrels. American red squirrels (*Tamiasciurus hudsonicus*) practice larder hoarding, constructing a large storehouse, or midden, of their pine cones and nuts. Their middens don't contribute as greatly to seed dispersal.

Researchers in the Greater Yellowstone Ecosystem discovered that the squirrels also contribute to the health of forests by potentially attracting other seed predators, thereby increasing forest biodiversity as they engineer the ecosystem.

Henry David Thoreau described his observations of squirrels using their caching behavior to plant forests in his essay "The Succession of Forest Trees," found in *Faith in a Seed*:

> On the 24th of September in 1857, as I was paddling down the Assabet in this town, I saw a red squirrel run along the bank under some herbage, with something large in its mouth. It stopped near the foot of a hemlock, within a couple of rods of me, and, hastily pawing a hole with its forefeet, dropped its booty into it, covered it up, and retreated part way up the trunk of the tree. . . . Digging there, I found two green pignuts joined together, with the thick husks on, buried about an inch and a half under the reddish soil of decayed hemlock leaves—just the right depth to plant it. In short, this squirrel was engaged in accomplishing two objects, to wit, laying up a store of winter food for itself and planting a hickory wood for all creation.

The squirrels and chipmunks Thoreau observed while paddling on the Assabet River and at Walden Pond have many descendants, and their caching behavior is still observed today by visitors to the historic site. Many visitors to Thoreau's celebrated pond and woodlands might miss, on their way to swim in his pond, the countless red oaks sprouting up under the trunks of eastern white pines amid clusters of acorn caps, but I was channeling the naturalist on my most recent visit on a hot summer day. The ardent work of the bushy-tailed rodents in the wood was as obvious to me as billboards on a paved highway.

But it isn't just the seemingly indiscriminate acorn planting that increases our carbon capture opportunities. In a 1993 paper titled "Tannins and Partial Consumption of Acorns: Implications for Dispersal of Oaks by Seed Predators," we find that "germination experiments revealed equal or greater germination frequencies for partially consumed acorns than for intact acorns." That's right: Partially intact acorns mean that squirrels took a bite and left the rest behind, and a tree still sprouted. The authors suggested that "higher tannin levels may render the apical portion less palatable, and thereby increase the probability of embryo survival after attack by seed consumers," also known as squirrels.

In another paper, published in 1986, "Grey Squirrel Food Preferences: The Effects of Tannin and Fat Concentration," biologists Peter D. Smallwood and Wm. David Peters from Ohio State University revealed that free-ranging eastern gray squirrels (*Sciurus carolinensis*) can determine which acorn species are more perishable. This knowledge can lead them to bury an acorn from a red oak, because it will last longer and germinate in the spring, while choosing to eat an acorn from a white oak (*Quercus alba*) because it will germinate earlier in autumn. But scientists, including Smallwood, revisited that finding two years later to investigate whether the size of the acorn made a difference. Was it just a little too large to eat on the spot and perhaps better to stash for later? Eating and caching behavior continues to be one of the more studied squirrel subjects, perhaps because there is so much riding on it, especially at a time when carbon capture is vital to our survival on this planet.

It's been found that many factors besides tannins and fat content go into nut selection. Squirrels consider nut size, nut mass, insect infestation, and food abundance, according to the authors of the landmark 2012 book *Squirrels of the World* by Richard Thorington Jr., John Koprowski, Michael Steele, and James Whatton. To this day, their book serves as the bible of species data for the family Sciuridae of the order Rodentia.

In 2017, in a study conducted at the University of California, Berkeley, scientists Mikel Delgado and Lucia Jacobs turned their attention to eastern fox squirrels (*Sciurus niger*) that buried their food in areas depending on what the food was, sort of like how we sort our own human pantries with baking supplies on one shelf and cans on another. What appears a random act is in fact quite determined, and those squirrel brains aren't as fuzzy as perceived.

Exploring the larder hoarding behavior of American red squirrels led to the discovery that it is intertwined with their social system. They are extremely territorial. These territories are evenly divided among squirrels, with mother squirrels often bequeathing their territory and middens to their daughters. Unlike the red squirrels, the scatter hoarding grays don't need such a territorial system. But these systems are not universal. Habitat and social structure can change in coniferous forests that are habitat for both reds and grays.

A squirrel above a busy rail trail in New York's Hudson Valley busily nips off oak branch tips filled with acorns as walkers and joggers dodge them below. The squirrel later darts past them on the trail with a nut in its mouth. This scene repeats during autumn as squirrels move about on their quest for food. For observers and researchers, it is yet another puzzling behavior we can add to the list of things we have yet to understand about squirrels.

"I can find no published scientific studies that reveal the reason(s) squirrels cut branch tips," wrote Joe Boggs, an assistant professor at Ohio State University. Boggs doesn't focus on squirrels. Insect pests and landscape problems are his specialty. Boggs addresses the various conjectures about this strange squirrel behavior, including that they "need to gnaw to wear down their ever-growing incisors." But, as he explains, this does not explain the cuts' apparent lack of gnaw marks or the seasonality of the branch nipping in late summer and early fall. Boggs leans toward another explanation. Since most of the nipped twigs include acorns, he surmises that it might be a harvesting technique used to gather the nuts at the smallest branch tips. This is often used by red squirrels to obtain cones.

This gray squirrel behavior plays out throughout the species' range. A forest health specialist with Wisconsin's Department of Natural Resources wrote about reports of oaks losing branch tips in 2018 that revealed tooth marks from squirrels and observations of the rodents clipping and dropping branches.

As often found in less remote deciduous forests, the researchers found that declines in seed dispersers, such as squirrel species, have led to simultaneous reductions in seed dispersal and seedling establishment, emphasizing the crucial role squirrels play in their forest ecosystems.

Studying potential long-term consequences of declining animal mutualists, including squirrels, on the structure and functioning of

Asian tropical forests is crucial for informing conservation strategies to safeguard these vital ecosystems and the species relying on them.

But there is also a need to explore the role of the native red squirrels in the forests of the United Kingdom and northern Italy. As the number of native squirrels decrease as they are replaced by invasive squirrels, how will this affect the health of European forests? So much of what we often discuss focuses on the loss of the iconic native squirrel species and the damage invasive squirrels have caused in native forests. Like reds in North America, Eurasian reds have their own niche. They influence the regeneration of pine woodlands in Europe. The reds are adapted to feed on native pine cones, while the grays feed more on deciduous, broad-leaved trees, such as oaks, in Europe's woodlands.

As with squirrels throughout the world, native species play an essential role in their ecosystems. Many native species of plants have a close ecological and evolutionary relationship with seed-dispersing native squirrels. Tree squirrels are vital seed predators and seed dispersers in both conifer and mixed forests in temperate and boreal biomes. These strong ecological relationships include the eastern fox squirrel (*Sciurus niger*) with longleaf pine (*Pinus palustris*) predation, Abert's squirrel (*Sciurus aberti*) preying on ponderosa pine (*Pinus ponderosa*), and Eurasian red squirrels (*Sciurus vulgaris*) with forest stands of Scotch pine (*Pinus sylvestris*), Corsican pine (*Pinus nigra*) in Belgium, and both Norway spruce (*Picea abies*) and Arolla pine (*Pinus cembra*) in the mixed conifer forests of the Italian Alps.

The Aleppo pine (*Pinus halepensis*) produces large, wingless seeds. The tree relies on nutcrackers and Eurasian red squirrels to disperse the seeds, and the squirrels depend on eating the green immature cones and the cones they scatter hoard for winter and spring. While the squirrels consume lots of seeds, the conifers have been shown to respond by producing nutritious seeds in tough cones that attract these animal partners, ensuring both seed survival and dispersal over wide areas.

These interactions illustrate the complex relationships between squirrels and conifers where squirrels act as both seed predators and seed dispersers, shaping forest dynamics. The protection of these interactions is vital. When connectivity biologist Jeff Gagnon started working with the Arizona Game and Fish Department to thin the ponderosa pines on Humphreys Peak from a forest fire perspective, he had to consider the squirrels' biology. "We would put little transmitters on the squirrels. I'd go out and track them to see where they were, measure their habitat to conserve those areas and keep them all connected. They are a keystone species in the forest plan," said Gagnon on a ride up to Arizona Snowbowl. The squirrels are prey for the forest's goshawks and seed dispersers for the pines.

But this isn't a story of just reds and grays or woodlands. Squirrels across the world influence their ecosystems.

Researchers traveled to Borneo to study the effects of squirrels and other wildlife on seed dispersal in the Indonesian rainforest, where around thirty squirrel species are found. Similar to the forests in Malaysia, the various squirrel species inhabit the four different layers of the rainforest by day and by night. For example, a selection of species, including the pale giant squirrel (*Ratufa affinis*), inhabit the top canopy layer and are active during the day, while the tufted ground squirrel (*Rheithrosciurus macrotis*) lives on the forest floor and is also active during the day. In contrast, the various flying squirrels each have a layer they inhabit at night, whether it be the canopy, understory, or middle layer.

In the end, hope lies with more conclusive population studies that will allow for greater understanding and possibly better management. Species distribution models (SDMs) are an important mode of population study. A recent study claimed that global Big Data models "are especially important for the many marginalized squirrel species . . . and the high number of endangered and data-deficient species in the world, specifically in tropical regions."

The study asserts that for the global squirrel species, "the habitat needs and ranges are widely unknown [and] not mutually agreed upon. . . . The tropics, particularly the forests of south and southeast Asia, are hotspots of squirrel diversity; however, this region generates the fewest scientific publications on squirrels."

A 2023 study addressed this precisely by including observations of multiple squirrel species, including three-striped ground squirrels, tufted ground squirrels, small-bodied Prevost's squirrels, pale giant squirrels, and red giant flying squirrels on the local forest community.

"I spend a significant amount of time alone, quiet and still, somewhere in a rainforest waiting, watching and writing down what animals are doing at certain fruiting trees," shared the author, Swapna Nelaballi, in an Instagram post in 2020 while in Indonesia's 108,000-hectare (approximately 266,900-acre) Gunung Palung National Park, where sunbears and pangolins live alongside the squirrel species.

Nelaballi and her fellow researcher interviewed local hunters along with observing tufted ground squirrels extracting heavy seeds from the extremely tough *Canarium* fruits "at remarkable speeds and consuming them onsite." Tufted ground squirrels, known locally as Reribu, have powerful jaws. The researchers also observed "three-striped ground squirrels visiting the fruiting *Canarium* tree." They observed that "three-striped ground squirrels were also capable of predation and secondary dispersal of these seeds, although we predominantly recorded caching, theft, and re-caching events by them. This discovery is significant, as it marks a novel finding; prior to our study, apart from few records of bearded pigs and giant squirrels no other mammal, apart from the tufted ground squirrel, had been reported to possess the capability to prey upon *Canarium* seeds. . . .

"Tufted ground squirrels exclusively cached *Canarium* sp. 2 seeds. The composition of the scatter-hoarding community we report mirrors that observed in other Southeast Asian forests." They found that "distant caches were mostly made by three-striped ground squirrels" moving the seeds away from the trees where they fell. "Rodents respond to seed dormancy, and tend to quickly move dormant seeds, take them farther, and re-cache them several times. Our observations support this, as we noted frequent and repeated re-caching of *Canarium* sp. 2 seeds by three-striped ground squirrels."

Their research in the remote rainforest was important in studying "the substantial impact of vertebrate seed predators, particularly large- and small-bodied generalists, on seed mortality and survival dynamics. . . . Lastly, our study suggests that small-bodied squirrels may play a crucial role in providing secondary dispersal services to plants facing limitations in dispersal agents."

While squirrels are beneficial to forest health, they also play a pivotal role in the health of an ecosystem's biodiversity as preferred prey for many wildlife species. Uinta ground squirrels (*Urocitellus armatus*) alone are a crucial food source for hawks, foxes, coyotes, and sandhill cranes, according to Rhea Cone, director of conservation at Utah's Swaner Preserve and EcoCenter. The ground squirrels, called *chislers* and Potguts, are native to western US meadows, pastures, and shrub-steppe habitats.

It's estimated that prairie dogs, the larger cousins of ground squirrels, contribute to the survival of 150 other species of their predators, such as endangered black-footed ferrets, coyotes, rattlesnakes, and the hawks that flew above the prairie dog colony I visited in Flagstaff, Arizona. "A ferret doesn't eat anything but a prairie dog, and if you don't have prairie dogs, that's really a bad day," said Jessica Simmons, urban wildlife planner for Arizona's Coconino County, on my visit.

Ernest Thompson Seton would have agreed with her. He wrote in *Lives of Game Animals*, "Now that the big Demon of Commerce has declared war on the Prairie-dog, that merry little simpleton of the Plains must go. In a few short years the tiny crater that erupted his annual families will be made no more, the older craters will be abandoned and crumble down to the level of the plain. And with the passing of the Prairie-dog, the Ferret, too, will pass."

While ground squirrels and prairie dogs can be easier prey than fast-moving arboreal squirrels, the nimble climbing squirrels are prey for many avian species, including long-eared owls (*Asio otis*), red-tailed hawks (*Buteo jamaicensis*), peregrine falcons (*Falco peregrinus*), and others in North America, and eagle owls (*Bubo bubo*) and Ural owls (*Strix uralensis*), among others, in Europe. Leopards (*Panthera pardus*), clouded leopards (*Neofelis nebulosa*), and even lion-tailed macaques (*Macaca silenus*) prey on arboreal giant squirrels despite their high perches in India's forests.

In addition to ground squirrels serving as a food source for many species, the intricate tunnels they dig provide shelter for wildlife, such as burrowing owls, and also aerate the soil, which promotes the health of native plants. Those plants are a crucial protein-rich, digestible vegetation for cattle. But there's the paradox. While ground squirrels, such as the Uinta, perform a strategic role in their ecosystem, those very holes and tunnels aerating the soil can be a nuisance to humans when they adversely affect landscaping and agriculture.

Our relationship with all squirrels, including arboreal species and ground species, is complex. Humans tend to put their interests first. The question of whether a peaceful coexistence can be obtained with any rodent, but specifically with squirrels, still needs to be determined.

*Chapter Two*

# In the World

*"The world is watching."*
—Joseph Biden, State of the Union Address, March 7, 2024

One morning years ago, a large stacked pile of long, papery cones mysteriously appeared under our spruce tree. It was as if the entire driveway had been tidied of the scattered cones overnight, like little elves had swept it clean while we slept, creating neat mounds under the trees that had released them. It took us all by surprise. But there weren't elves or human gardeners at work at our house; there were red squirrels.

The small American red squirrel that visited my spruce was one of the three species of tree squirrels classified in the genus *Tamiasciurus*, from the Greek ταμίας, *tamías*, meaning "steward, dispenser," and the Greek for *squirrel*. Scientific names, unlike common names, don't often reflect the species' behavior. *Tamiasciurus hudsonicus*, for example, indicates only where German naturalist Johann Christian Polycarp Erxleben first documented the chattering mammal, in 1771 in North America's Hudson Bay, where it was then placed in the genus among other pine squirrels. John James Audubon referred to it as the Hudson's Bay Squirrel, Chickaree, or Red Squirrel when he portrayed the squirrel in *The Viviparous Quadrupeds of North America [1845–1849]*.

But most call the squirrel by its common names, including chickaree, barking squirrel, pine squirrel, and spruce squirrel, which denote much of the squirrel's vocal behavior. Red squirrels are highly territorial and chatter loudly at anyone or anything passing by their conifer territory. Throughout their lives, American red squirrels (*Tamiasciurus hudsonicus*) perch in pine trees and drop their scraps, the scales and woody bracts of cones, at the base of the tree.

Red squirrels are just one of the squirrel species that live in New York. The others include the common eastern gray squirrel (*Sciurus carolinensis*); two species of elusive, mostly nocturnal flying squirrels, the southern (*Glaucomys volans*) and northern (*Glaucomys sabrinus*) flying squirrels; and other cousins in the squirrel family, the eastern chipmunk (*Tamias striatus*) and the woodchuck (*Marmota monax*), also known as the groundhog or whistle pig. The eastern fox squirrel

(*Sciurus niger*), although primarily a Midwestern species, also has a small population in the western part of the state.

A fungal disease of spruce trees called *Rhizosphaera* needle cast and environmental stress had damaged our New York Hudson Valley spruce trees, forcing us to remove three of them. We resisted the inevitable departure for far too long, hoping in vain that they would recover. Their loss left a gap in our yard and diminished an important source of cones for our red squirrel neighbors.

That pile of cones at the base of my spruce was a carefully gathered midden. These structures, created by squirrels under their favorite tree, can span fifteen to thirty feet in diameter and rise up to four feet tall. Each midden also included the green cones the squirrels had cut from the trees and placed for winter storage. These middens play a role in protecting biodiversity.

My observation of this one common squirrel was just one among millions of observations around the world of hundreds of squirrel species. Those squirrel species are some of the more than 65 living in North America among the 294 squirrel species found naturally worldwide, including such variations of the marvels of evolution as pygmy, dwarf, giant, flying, giant flying, beautiful, sun, groove-toothed, palm, long-nosed, rock, red-checked, American red, tree, striped, and American bush squirrels; chipmunks; ground squirrels; prairie dogs; and the largest, the marmots. In addition, a groundhog famously predicts the beginning of spring, rounding out the full squirrel extended family. It has been said that "the sun never sets on the Sciuridae!"

It can be challenging to keep all these squirrel species and their statuses straight. Most of us are familiar solely with the ones we see regularly, and we assume, mistakenly, that all squirrels are abundant and spend their days scurrying up and down the trunks of trees. That's a partial view, a sliver of our understanding. We must dig into the entire group and its biology to obtain a more accurate view of what constitutes a squirrel.

Squirrels belong to the order Rodentia, and within that order they form the family Sciuridae. As a group, the Sciuridae are diverse and make up 294 species and 51 genera, broken into five subfamilies (Ratufinae, Sciurillinae, Sciurinae, Xerinae, and Callosciurinae). These subfamily names point to the distinctive characteristics of these squirrels, as with Callosciurinae, the Asiatic squirrels, containing over 60 species named after the genus *Callosciurus*, meaning "beautiful squirrel."

Of all the species in the Sciuridae group, 122 are tree squirrels, and 62 are ground squirrels. With some exceptions, squirrels are characterized as small, bushy-tailed rodents with slender bodies and large eyes, but the variations can be dramatic. Just look at two squirrels from Japan to see this dramatic range: the Japanese pygmy dwarf squirrel (*Pteromys momonga*), weighing less than ten ounces, and the eight-pound Japanese giant flying squirrel (*Petaurista leucogenys*).

Size isn't the single characteristic with such a wide range among squirrel species. Squirrel fur is short and soft but varies in color and thickness among species. Squirrels can be gray, yellow, red, brown, black, or even white, and striped or not striped. Members of the tribe Marmotini—a tribe is a taxonomic rank between subfamily and genus—display a remarkable variety of colors and patterns. Tail length and color also come into play when identifying squirrels. Squirrels of the subfamily Ratufinae, for example, have extra-long tails.

Squirrels' behavior and habitat also contribute to their physical characteristics. Arboreal, or tree, squirrels have long, bushy tails, large ears, and sharp claws for tree climbing. Flying squirrel species are defined by a furred membrane called a patagium that extends between their ankles and wrists and allows them to glide from tree to tree. Ground squirrels have more robust bodies and less bushy tails than tree squirrels. They also have short, sturdy forelimbs to assist them with digging burrows. Whether gray, red, brown, or black, all squirrels have the same skull architecture, including being short with a short front of the skull, called the rostrum, and having an arched profile.

Squirrels are omnivores, eating everything from nuts to plants to meat. While they mainly eat tree seeds and fruit, they'll also eat insects, fungi, buds, shoots, flowers, bark, smaller rodents, bird eggs, caterpillars, snakes, and, much to the dismay of bird-watchers, even small birds.

This ecological role was echoed on a larger scale after Washington state's Mount Saint Helens erupted in 1980, covering over twenty thousand miles with 540 million tons of ash. Three years later, scientists tested whether burrowing rodents could jump-start recovery of the devastated landscape. They air-dropped endemic pocket gophers—a relative of ground squirrels and prairie dogs—into the area. The gophers' digging behavior was a powerful ecological tool, enhancing soil structure and microbial activity. Six years later, the researchers recorded forty thousand plants flourishing in the spot. In contrast, the surrounding area was still very barren.

Each squirrel species requires a specific habitat for its survival. Gray and fox squirrels, for example, require mid- to late-succession forests for their success. The most productive habitats have a variety of tree species. They will, however, forage in some early successional stages, such as clear-cuts and crop fields. Both species are found on bluffs, riverbanks, and bottomlands. As with many squirrel species, a suitable habitat must contain food sources through all seasons.

Nocturnal flying squirrels have their own habitat needs. Southern flying squirrels often nest in dead beech trees and red oaks. They also make their homes in abandoned bird nests or the dreys (nests) of other squirrels, in woodpecker holes, and in available attics. "You don't encounter them like you would your typical gray squirrel . . . your daytime active squirrels. But they're probably as common if not more common than gray squirrels in some areas," said retired Rhode Island Department of Environmental Management wildlife biologist Charlie Brown about the southern flying squirrels.

Renae Swayser, an environmental educator for Pennsylvania's Wildlands Conservancy, knows this all too well. In February 2024, on a cold Saturday morning, she led a hike to check the squirrel nesting boxes at Thomas Darling Preserve in the Poconos of Pennsylvania, hoping to find endangered northern flying squirrels. The fifty or so nesting boxes had been placed in trees about twenty feet off the ground, and climbing a ladder was necessary to check them. After knocking on them all, she found three southern flying squirrels in one box, but no northern flying squirrels. "Southern flying squirrels are typically the ones that a lot of people [see] if they have cabins in the Poconos, or, you know, a house that they might not visit that often or like a summer house, those are the ones that you find in your attic," she said. While southern flying squirrels might be as common as grays in some areas, northern flying squirrels are endangered. According to the Pennsylvania Game

Commission, an extensive population analysis from 2003 through 2007 found barely thirty-three northern flying squirrels, mainly in the Poconos. Conducting a population study that finds northern flying squirrels, such as the one Swayser is involved with, might move the game commission to develop a plan to help better manage the squirrel's forest habitat.

The white-bellied northern flying squirrel, the other of the two North American flying squirrel species, resides in deciduous and mixed woods from southeastern Canada to Florida. Forest and agricultural tree habitats impact other species of squirrels besides flying squirrels. Woodland habitat management is crucial to preserving our squirrel biodiversity. A similar example is a recent study of threatened Persian squirrels (*Sciurus anomalus*) on the Greek island of Lesvos, which showed that modern olive tree pruning practices jeopardize the squirrel population. The single tree squirrel in the region is the Persian squirrel, with a grayish-brown color on its back, a yellowish-brown color on its belly, and a yellow-brown to deep red tail. It is considered a valuable keystone species on Lesvos, where it lives in centennial olive groves.

All squirrels, including those two flying squirrel species, are active and crucial in their ecosystems. Marmots, prairie dogs, and other ground species dig burrows in shortgrass habitats that become used by many different species. All serve as prey for various predators, from hawks to humans. For example, northern flying squirrels (*Glaucomys sabrinus*) comprise at least 50 percent of the prey in the diets of the critically endangered spotted owl found across forests in the Pacific Northwest states of Oregon and Washington. The loss of this species would significantly impact the population of endangered owls.

Those ground squirrels include species most don't consider to be squirrels. The little chipmunk filling its pouches and appearing in children's *Chip and Dale* cartoons is an example. It is just one of the 25 species of squirrel in the *Tamias* genus, one of the 50 squirrel genera. All chipmunks are squirrels, but not all squirrels are chipmunks. All but one of those 25 species of chipmunks are native to North America. The Siberian chipmunk is the exception and lives in Asia.

Prairie dogs, another ground squirrel species, elicit a wide response from their human neighbors. According to one study, "the prairie dog has been cussed, discussed, protected, exploited, credited with doing so many good things, and accused of being completely bad." Some call them prairie rats, and others consider them a keystone species. Colonies of prairie dogs, once seen as agricultural pests by the United States government, were poisoned on behalf of agricultural interests. It wasn't until their obligate predator, the black-footed ferrets that once numbered in the tens of thousands, were listed under the Endangered Species Act of 1973 that prairie dog habitat was protected.

Today, prairie dogs are considered a species of greatest conservation need in the four states they inhabit. Drought and plague still hamper prairie dog populations, according to doctoral student Emily Renn. She has worked in Arizona to relocate colonies of Gunnison's prairie dog (*Cynomys gunnisoni*) from urban areas that are threatened by development or destruction to Petrified Forest National Park and other areas with hopes of reestablishing the extirpated wildland colonies. Renn serves as the translocation/coexistence coordinator for Habitat Harmony, an organization that coordinated with the Arizona Game and Fish Department to write the 2015 translocation protocol for Gunnison's prairie dogs in Arizona and authored the 2018 guide to nonlethal management of Gunnison's prairie dogs.

While drought and plague are threats, prairie dogs have other challenges. "They only have one litter of pups, usually three to five pups in spring," Renn told me. "Unlike other rodents that breed regularly, prairie dogs only breed on one day of the whole year. Each individual female is only receptive on one day." If those weren't enough challenges for prairie dog survival, Renn claims that hunter surveys between 2000 and 2006 reported that there were anywhere from 30,000 to 94,000 prairie dogs shot in Arizona each year. Those numbers were acquired through voluntary hunting reporting, so they probably greatly underrepresent what's actually happening, according to Renn. While tribes do shoot prairie dogs for food during the fall, Renn claims that the number has a "small impact compared to target shooting."

Back in my yard, I haven't seen those small reds since we removed the spruce trees. Their absence has left a noticeable void in my daily life. My driveway was messy without them eating and gathering cones for their midden. Pine cones from the remaining trees were strewn everywhere. Nothing or no one tidied them up. The snowplow pushed them to places they'd never been. More important, my observation showed that our squirrel biodiversity had decreased, a fact that I find deeply troubling.

At most, the survivorship of squirrel juveniles is about 22 percent, meaning that many baby squirrels become prey to other species. Those predators might have gone elsewhere without the squirrels inhabiting my spruce habitat. Red squirrels are avian nest predators on the other side of the predator–prey relationship. When squirrels are in abundance, nestling survivorship may be low. Removing the squirrels from the ecosystem has consequences. I'm not sure what happened to the reds at my home after the spruce trees came down, but I know I do miss them. And I can't stop wondering about the health of the red squirrel population in an area that is losing so many spruce trees beyond the ones at my house. Currently, no studies are exploring the red squirrel population where I live. As with Renn's experience in Arizona, New York relies on hunter records for squirrel population information. Without accurate studies, it is difficult to know the true state of the population, just as I can't measure the state of red squirrels by what is happening at my house. This underscores the urgent need for scientific studies to provide a more complete picture.

Squirrels, though common, play a crucial role in our ecosystem. We often don't notice their presence until they are gone. And mine were gone. I didn't notice their disappearance until the cones from my remaining spruce trees flooded my driveway. Although they have lived beside us forever, squirrels are rarely considered beyond National Squirrel Appreciation Day or studies by wildlife biologists.

Biologist Thaddeus McRae's comment that squirrels are background rings true. Are squirrels now just white noise in the swelling soundtrack of our noisy lives? As headlines scream of wildfires, floods, and a planet unraveling under the weight of human-induced climate collapse, who still pauses to wonder about the fate of squirrels? It seems entirely possible that these familiar creatures—and many others—are vanishing quietly, drowned out by the louder crises we face and the accelerating loss of life around us.

Given the remarkably minimal scientific interest in them, it's clear that we need to focus more scientific scrutiny on squirrels before we lose species most stressed by humans and climate change. Worldwide, squirrel species might be the most mysterious creatures we see every day. In this era of worldwide biodiversity loss, it's imperative to bring squirrels out of the background and into our attention, and take action to protect them.

Imagine an annual accounting, like the State of the State, each year. A similar event for our global squirrels would present a clear view of how our squirrels are faring, including which ones are struggling and which are thriving.

A similar event, the International Colloquium on Arboreal Squirrels, does exist, but unfortunately, it doesn't include all squirrel species and doesn't occur annually. The last one was held before the COVID-19 pandemic, interrupting the schedule of having it every three years. Without it, we're lacking vital information.

"From deer stands, I have often watched squirrels rustle about in the leaves beneath me, or heard them in the trees above and behind me as they scold me as an intruder, so I believe that our countryside has no squirrel deficiency," wrote Steve Gilliland from Kansas in 2024. Is Gilliland right—if we see squirrels around us, should we assume there is an abundance? Or should we believe there isn't if we don't, as in my case?

This thinking pattern corresponds to the way many people think about climate change. If it is cold at my house, does it mean that Earth isn't warming? Remember when Senator James Inhofe from Oklahoma brought a snowball into the US Senate in 2015 to demonstrate that the globe is not warming? "I asked the chair, do you know what this is? It's a snowball just from outside here. So it's very, very cold out. Very unseasonable." Are our casual squirrel observations just as tone-deaf?

I recognized that the lack of red squirrels at my house didn't indicate a decrease in the population locally or regionally. I'd have to search for information from the experts, the state agency tasked with conservation, to test my hypothesis that the struggling and dying spruce trees in my area might adversely affect the local red squirrel population.

If my hypothesis was correct, would the local annual squirrel hunt be allowed to continue? Indeed, if the population dipped, I believed that it would certainly be curtailed.

However, the experts at the New York State Department of Environmental Conservation did not have the answers I needed. As mentioned, I learned that the department, as in Arizona, did not study the population; the agency relied solely on hunters for the status of the state's squirrel populations. It had nothing unusual to report or any completed scientific studies.

In addition, the local squirrel hunt, the Squirrel Scramble, was indeed proceeding. It advertised that hunters would receive extra points for killing red or black squirrels. Black squirrels are a melanistic subgroup of squirrels with black fur, not a separate squirrel species. They are a rarer color variation of the eastern gray squirrel.

At that time, I didn't know that the local squirrel-hunting contest would be the last one in New York state. It wasn't ending because of concern about the squirrel population. Instead, after continued pleas from animal protection and conservation organizations and legislation passed by the state legislature, New York's governor, Kathy Hochul, signed legislation S.4099/A.2917 on December 22, 2023, to protect the state's wildlife by making hunting contests, competitions, tournaments, and derbies that allowed for the taking of large numbers of wildlife unlawful.

In a press release, Governor Hochul said, "Protecting wildlife is critical to fostering the integrity and resilience of our environment and outdoor recreation economy. This legislation establishes strong safeguards for our state's precious wildlife species and protects our important fishing and hunting traditions." Humane World for Animals estimates that at least 57,000 animals, many of them, like squirrels, considered a "nuisance," are killed yearly in wildlife killing contests and competitions. With more states eliminating hunting contests, that total number is down from more than 60,000.

Although the legislation in New York excludes contests for hunting white-tailed deer, turkeys, and bears and for fishing contests, it will affect the populations of other species, including squirrels. "Today is a win for every animal that was previously targeted by these cruel contests," said Regan Downey, director of education at the Wolf Conservation

Center. "We applaud Gov. Hochul's decision to sign A.2917/S.4099 into law. New Yorkers value humane and science-based approaches to wildlife management, and we are thankful to finally have a policy that reflects these values in our backyard. Killing contests have no place in the 21st century, nor do they have a place in New York."

The new law makes it illegal "for an individual to organize, sponsor, conduct, promote or participate in any contest, competition, tournament or derby with the objective of taking or hunting wildlife for prizes, inducement or entertainment. Any wildlife killed during these activities become the property of the New York State Department of Environmental Conservation."

Amid backlash from those who feel these hunts are a popular and family-friendly way of introducing their children to hunting, New York state became the tenth state to stop the slaughter of wildlife for cash and prizes, following a recent Oregon ban of contests on state land. California was the first state in the country to ban wildlife contests, in 2014. Illinois and New Jersey were following New York's lead in 2024 to ban hunting contests through legislation rather than state agency rules. New Jersey state representative Anna Moeller, the sponsor of the New Jersey bill, said to a reporter from the *New Hampshire Bulletin*, "We support hunting that's done in a sustainable and responsible way. When you're wiping out large numbers of animals at a time, you're creating an imbalance, and oftentimes we find there's harmful consequences from being so reckless."

In March 2024, New Jersey introduced N.J. Admin. Code § 7:25-5.22, stating, "No person shall have in possession, kill, attempt to take, hunt for, pursue, shoot, shoot at, trap, or attempt to trap any wild mammal or wild birds unless an open season for the taking of such birds or mammals has been declared by the New Jersey Fish and Game Laws or Code and then only during the respective open seasons." The introduced legislation can be viewed as an animal bill of rights. It prohibits any person from organizing, sponsoring, promoting, conducting, or participating in a competitive event at which participants harass or take covered wildlife, except in conjunction with an authorized field day event.

On January 23, 2024, a public radio program focused on a popular hunting contest in New York's North Country region. In a poll conducted by *North Country Now*, 73 percent opposed the new law. The Norfolk Rod and Gun Club held a rabbit and squirrel hunt for the last time in January. Many hunting clubs, like the Norfolk Rod and Gun Club, rely on these events to sustain their organization's membership. Participants in the contest who killed at least one squirrel were entered in a raffle. Twelve-year-old Gabby Haggett won a scope for shooting and killing a squirrel in the competition. She said she would take the scope into the woods near her family's farm. "I like to go out there and just bring my gun and see what I can find out there so then Dad'll let me have this scope while I go out." This contest was probably the last competitive squirrel hunt for the club before the law goes into effect.

"These contests result in the wasteful slaughter of large numbers of a particular species in an area," said Deborah Glick, chair of the state's Environmental Conservation Committee and sponsor of the new state law. She added, "Nothing in this bill prevents people from hunting under the regulations that exist from the Department of Environmental Conservation, and so, for those who augment their family's food supply, nothing in this prevents that."

In a press release, Glick said, "It is shocking that in New York, dozens of these barbarous, unsporting contests take place each year to kill the largest number of certain species of wildlife. These killing contests serve no conservation or scientifically backed ecological purpose and encourage senseless brutality. I applaud Governor Hochul for signing this legislation and ending this inhumane practice while protecting a farmer, rancher, or other New Yorkers' right to safeguard companion animals and livestock from nuisance animals through DEC regulations. The wildlife of New York is a natural resource that should be protected, not brutally killed for cash."

While the hunting contest ban is a win for New York state's squirrel population, squirrel-hunting contests continue throughout the United States. Like the former contest in Germantown, New York, known as the Squirrel Scramble, these hunts, as Glick said, ask contestants to kill as many squirrels as they can in a limited time frame. Contestants receive prizes for squirrel weights and numbers killed. This hunting practice differs from traditional hunting for squirrels during regulated hunting seasons because contests are not grounded in scientific knowledge as a management strategy. The contests make no meaningful contribution to conservation but do draw revenue through license sales.

In contrast, despite animal rights protests, in 2023 the Michigan Natural Resources Commission expanded the animals considered nuisance to include foxes, red squirrels, gray squirrels, and ground squirrels, allowing them to be killed and trapped without a permit.

Some people, such as Josh Donaldson, known as Earthdrop on social media, question the idea of calling certain animals "pests," pointing out that it's not the animals' fault humans have built homes on their natural habitats. From this perspective, these creatures are simply trying to survive like anyone else, just trying to make it through another day.

Treating squirrels as disposable pests and targets worth no more than prize money harms the essence of hunting and the social relationship between hunters and nonhunters, especially when there isn't definitive knowledge of the health of the species' population.

Professional golfer Brian Harman gave an interview in 2020 in which he opened up to *Golfweek* about his hunting values. "It (the lesson from his father) built a deep respect for animals. What they provide. They are a renewable resource. Being able to know where your meat comes from is important to me, being able to take care of the animal once you've killed it shows immense respect to the animal." Harman continued by stating that he is "not a fan of people who kill for sport."

But are animals indeed a renewable resource, as Harman declared? *Renewable* is defined by the *Oxford English Dictionary* as "not depleted when used" and "capable of being renewed." Is the term *ubiquitous* synonymous with *renewable*? According to the Texas Parks and Wildlife Department, wildlife is a renewable resource "which, properly managed, replenishes [itself] indefinitely." The question then becomes, Are squirrel populations adequately managed? According to the same Texas hunting guide, this belief structure contrasts with "wildlife preservation, which is the saving of natural resources without any consumptive use of them." Without knowing the full status of a species, it seems impossible to manage it properly.

In 1905, Louisiana's *Lafayette Advertiser* warned that since American alligators helped prevent flooding by preying on muskrats that created holes in levees, there would be dangerous ramifications if alligator hunting continued. Despite raising the alarm and limited regulations, hunting of the ubiquitous gators continued in and out of the state at a fevered rate. The state's Department of Wildlife and Fisheries estimated that "approximately 3.5 million Louisiana alligators were killed" for their skins from 1880 to 1933. Although commercial and recreational gator hunting was finally curtailed in 1962 to protect the species, the decreased population from hunting still led to the alligator's listing as one of the first species under the newly created Endangered Species Act of 1973. American alligators were able to rebound after protections were established. But that's not always the case, as we saw when the Lacey Act of 1900 failed to save the ubiquitous passenger pigeon, which went extinct in 1914.

Arkansas is one of the states that have ongoing hunting contests. The Arkansas Game and Fish Commission's 2023 statewide Big Squirrel Challenge is billed as a squirrel-hunting celebration that rewards hunters who present the weightiest squirrels. The state is home to resident fox and gray squirrels, with fox squirrels a little larger. Like other wildlife-hunting contests, the hunt has the potential to impact the population at the end of the day. What is the current state of the squirrel population in Arkansas? Is the hunt proceeding without a full accounting?

The US Fish and Wildlife Service solely lists wildlife species on its official list of federal endangered and threatened species after receiving petitions from state governments or outside agencies such as universities. The federal agency conducts its population analysis, with the concern rising first from other studies. But what if those other studies aren't occurring?

Two months after January's Squirrel Appreciation Day, a radio station on the Isle of Wight shared locations to see native red squirrels over the Easter break, when the favorite rodents are out and about getting their dreys spiffed up for breeding and caring for newborn kits. The Isle of Wight, the largest English island, has become one of the few places in England with a resident red squirrel population. One of the best spots on the southern island for squirrel spotting is a group of reserves, the Alverstone Mead Nature Reserve complex, home to a thriving squirrel community free from eastern gray interlopers. While this island has become a refuge for this native population, islands, often isolated and scattered, can also pose a conservation threat because wildlife cannot expand or shift their habitat. These locations are considered in conservation to be high-risk regions, according to "Squirrels on Islands: The Effect of a 'Laissez-Faire' Approach from Governments and Their Responsible Entities on the Marginalization and Extinction in Extremely Restricted Habitats," a chapter in *Sustainable Squirrel Conservation*, authored by Netherlands researcher Moriz Steiner and Falk Huettmann from the University of Alaska. In this case, the island has allowed the reds to live without competition from the invasive squirrels. In Essex, England, Mersea Island also boasts a thriving red squirrel population after a local community worked to reintroduce the threatened species to the British island. The group was inspired by the surviving populations on the Isle of Wight, Anglesey in Wales, and Brownsea Island. This sounds great for the red squirrels, but islands face their challenges on this warming planet.

The Isle of Wight and the other islands in the British Isles are no exception. Summers are becoming hotter and drier, which will affect water resources, crop growth and harvests, and the risk of wildfires, all of which will affect those native squirrels. Winters will be milder and wetter, increasing the likelihood of flooding and landslips. Sea level rises will affect availability of land. All of these challenges are becoming common to all islands. And while the Isle of Wight might not suffer from a laissez-faire government approach, other islands do.

Relegating our squirrel populations to islands that either are surrounded by water or are islands of fragmented habitat within built environments does not bode well for global squirrel populations.

While we are still determining the state of our global squirrel populations, I still don't know the status of my local red squirrel population. However, I do know that we can learn more about our squirrel populations only by observing them in all habitats. Observing squirrels is perhaps what we all, scientists included, do best.

President Joseph Biden said, "The world is watching," in his 2024 State of the Union Address. Although he certainly wasn't speaking about the global squirrel population, the world is watching and has been watching squirrels for millennia. Those observations have found their way into our stories and our science. And like any State of the Union Address phrase, *the world is watching* is meant to be a wake-up call.

*Chapter Three*

# As a Cultural Icon

*"Humans and nature construct one another."*
—Alexander Wilson

I sat alone one afternoon enjoying the crisp air of a sun-filled autumn day oceanside in Maine. My copy of Rachel Carson's *Silent Spring* rested beside me, my camera in my hand, when a red squirrel's movement in a nearby tree caught my eye. With no distractions and the time to enjoy the solitude of the setting, I watched this busy little classic American red scramble up and down the bark of a hemlock trunk. It would descend, pick up a cone, and carry it back up the trunk to a branch, where it would hold it with its two front paws and proceed to nibble it like an ear of corn, stripping away the scales to get to the nutritious seeds within. Up and down, up and down. Cone after cone. Scaly bits falling below. A meditative cadence. As I sat there, the squirrel became more and more comfortable with me edging closer with my camera. I can't say how long I sat there, but I recall the sun setting and the horizon turning brilliant orange as I took many photographs of my busy squirrel subject.

Native North American red squirrels are about twelve inches long, and their fur is not always red. They can also have grayish or rust-colored fur with a white belly. They can even have a black stripe on the side of their body. This one was tiny, red, and very industrious.

The squirrel show was local and free. Thank you, little red.

"No other animal looks so much like eagerness incarnate as a squirrel standing up on its hind legs, sniffing, erecting its ears, hands at its chest, eyes wide and wet," wrote naturalist Diane Ackerman in her 1995 *National Geographic* article "In Praise of Squirrels."

She described squirrel watching as a "front-row seat at one of life's little operas." She couldn't have been more correct. Watching squirrels in their daily pursuits is entertaining. Not only is it easy and cheap; most people can see a squirrel and enjoy the show by just glancing out a window or taking a walk in a local park.

Her observations are joined by those of Annie Dillard, who wrote in *Pilgrim at Tinker Creek*, "I sit on a fallen trunk in the shade and watch the squirrels in the sun."

Squirrel observations abound in all cultures. They've been immortalized for millennia in our myths and legends. Our humble and familiar member of the rodent family figures prominently in Native American, Norse, and Hindu tales as a chatterbox, a gossip, a symbol of resourcefulness, and a mystical, powerful being that can bridge the worlds of the living and the dead.

While squirrels feature in many Indigenous stories in North America, a legend orally shared from the Haudenosaunee Confederacy is perhaps one of the most well-known. It credits squirrels with sparking humans to tap and evaporate the sap that runs through maple trees. That sap becomes the sweet syrup we enjoy on our pancakes and waffles.

Canadian naturalist Frère Marie-Victorin described how red squirrels (*Tamiasciurus hudsonicus*) helped inspire early maple sugaring. The legend explains how one spring day, a young man watched a squirrel climb a maple tree, bite into a branch, and drink the tree's sap. Eventually, the man slashed the tree with his knife, mimicking the squirrel, and collected the sap.

The story of the squirrel and the maple sap corresponds to Western scientific studies and further observations. Bernd Heinrich's 1992 paper in the *Journal of Mammalogy*, titled "Maple Sugaring by Red Squirrels," explores the observation of chisel-like grooves systematically made by American red squirrels on western Maine sugar maples (*Acer saccharum*). The tooth marks left by those sugaring red squirrels were observed at twenty-two other sites in Maine and Vermont. The study exemplifies how traditional ecological knowledge and Western science often align.

This traditional story is just one of many squirrel tales that appear in our cultural history. First-century CE Roman naturalist Pliny the Elder wrote of his squirrel observations in his native Italy in *Natural History*: "Squirrels also foresee a storm, and stop up their holes to windward in advance, opening doorways on the other side; moreover their own exceptionally bushy tail serves them as a covering."

Pliny's observation was correct; squirrels do use their bushy tail as a covering. In fact, the scientific name for the squirrel family is derived from Greek and Latin words for shade and tail, meaning shadow-tailed.

The noun *sciurus* is a Latin derivative of the Greek word *skiorus*, which is a compound word for *skiá*, "shade," and *ourá*, "tail."

We watch squirrels flick their bushy tails in annoyance, wrap their tails around themselves for warmth, and use them for balance like the most agile tightrope walkers. Perhaps their attractive, bushy tail separates how we feel about them from how we think about hairless-tailed rats and mice.

Emily Dickinson is just one of our many modern poets who have written squirrels into their poetry, in poems like "Nature is what we see—": "'Nature'" is what we see— / The Hill—the Afternoon— / Squirrel—Eclipse—the Bumble bee—" and, from "What shall I do when the Summer troubles," phrases like "Oh, when the Squirrel fills His Pockets / And the Berries stare / How can I bear their jocund Faces / Thou from Here, so far?" Perhaps her poems celebrated the squirrels outside her Amherst window or she remembered them scurrying around the nearby Mount Holyoke College campus while she was a student.

Other poets and poems memorializing squirrels include Joan Murray's "To appreciate Squirrels" and Mary Oliver's "Making the House Ready for the Lord." Over four hundred *Washington Post* readers entered the 2019 Squirrel Haiku Contest during the paper's ninth annual Squirrel Week. Post Metro columnist John Kelly, the founder of Squirrel Week, who helped judge the contest, said, "The submissions convinced me that these young poets spent a lot of time looking at squirrels, which is the only way you can learn to appreciate something." Kelly is right; observation is key.

Far away from the woodlands of North America, another red squirrel appears in the ancient tales of Norse mythology and Scandinavia. Ratatoskr is the nimble red squirrel inhabiting those Scandinavian myths and woodlands. The stories are most likely inspired by the native Eurasian red squirrel (*Sciurus vulgaris*), a different species from its North American red cousin. Unlike American red squirrels, which do not have tufted ears, Eurasian or European (*S. vulgaris*) red squirrels do. That legendary red squirrel of Norse mythology is gifted with the ability to communicate between the roots of Yggdrasil, the Norse Tree of Life, a massive ash where the serpent Níðhöggr lives, and the tree's canopy, where the eagle Hræsvelgr resides, carrying messages between the two and to all the realms.

In the *Poetic Edda*, the modern name for a collection of Old Norse poems written in the thirteenth century, Ratatoskr appears:

> *Ratatoskr is the squirrel who there shall run*
> *On the ash-tree Yggdrasil;*
> *From above the words of the eagle he bears,*
> *And tells them to Níðhöggr beneath.*

The *Prose Edda*, another Old Norse verse, written in the same century, also includes Ratatoskr, in which the troublesome mythological squirrel is described as running up and down the ash tree telling slanderous gossip and provoking both the eagle above and snake dragon below.

Squirrel stories also appear in India. The tale of the squirrel in the ancient Indian Hindu epic poem the *Ramayana* tells of a small squirrel collecting pebbles. Lord Rama needed to build a bridge across the sea to Lanka to rescue Sita. While Lord Hanuman and the monkeys carried large boulders and rocks, the small squirrels carried pebbles. A monkey carrying a heavy stone shouted at the little squirrel, "You little thing! You're in my way. I almost fell. What are you doing here?" However, the squirrel's efforts proved vital in the bridge construction by filling in the gaps between the large boulders that formed the bridge. The small pebbles made the entire structure stronger. It is said that Rama thanked the squirrel for its hard work and stroked its back, explaining the three dark stripes on the back of palm squirrels (*Funambulus palmarum*) found in South India today, where the species is considered sacred and not to be harmed.

Today, a fifteen-foot steel statue of a squirrel stands in the atrium of India's Ayodhya Dham Junction railway station. Bengaluru-based sculptor Kalyan Rathore created it to celebrate that epic Hindu story. Hearing that story during his childhood inspired him. "I want people, especially children, to take home the message that it doesn't matter how little you can contribute as long as you contribute . . . that's what matters."

In North America stands another red squirrel sculpture in front of the Galloway Station Museum in Alberta, Canada. Eddie is the friendly mascot of the town of Edison. The current statue, Eddie Jr., crafted of fiberglass, replaced the first cement sculpture, Eddie Sr., which stood for thirty years until 2011. Eddie Jr. is perched on a log and holds a bronze pine cone instead of an acorn because there are more pine cones than acorns in the area.

Heading south to Texas, the Seguin Squirrel Trail provides fun for residents and artists who enjoy the public art program's decorated fiberglass squirrel sculptures all around the community, distributed in 2023. An interactive map leads visitors to the spots to locate River Guardian Squirrel, Flora, Curiosity, and Chance, among the other squirrel sculptures around the community that individuals and organizations sponsor.

The village of Glendale, Ohio, has its own fiberglass squirrel sculptures to celebrate the large population of its much-loved black squirrels. The original 25 five-foot-high sculptures were presented at the village's sesquicentennial celebration in 2005. Today, there are still 13 of those squirrels in the village. As in Seguin, visitors can access an interactive map online for the self-guided squirrel statue tour.

And Marysville, Kansas, boasts a black squirrel as the city's mascot and has over fifty squirrel sculptures in a collection titled "Black Squirrels on Parade." The city's famous black squirrels originated with two squirrels that escaped from a cage in the McMahon Carnival in 1912. The melanistic subgroup of gray squirrels became protected and the official city mascot in 1972.

Regardless of location, whether in ancient tales or current culture, stories of powerful, industrious squirrels have been transformed and adapted into many forms, including sculptures, fun and frolicking cartoons, and even superhero comic book characters. For example, readers will find a villainous female squirrel god from Asgard in Marvel's *The Unbeatable Squirrel Girl* comic book series and *Marvel's Squirrel Girl: The Unbeatable Radio Show*. But just as with actual small squirrels, the comic features a character more often overlooked and insignificant. "The initial approach was to find someone really obscure. We also had some villains that might be natural enemies of squirrels, like Princess

Python," explained Dean Hale, coauthor of the middle-grade graphic novel. "It was to mirror the idea of how Squirrel Girl is overlooked because most villains look at her like she is sweet and insignificant, and that's to their detriment." But the Hales also chose to give their superheroine traits associated with real squirrels, such as agility and speed, combined with proportional strength and the characteristic bushy tail.

While Doreen Green, also known as Squirrel Girl, might be one of Marvel's more courageous and powerful characters, she is sometimes viewed as a joke. On the other hand, she is also a much-beloved character for many fans and enjoys a cult following online. But Squirrel Girl isn't the sole character inspired by the squirrels of Norse mythology. Comics fans can enjoy a scurry, or group, of adventurous Viking squirrels across the pages of the *Sons of Ashgard* saga.

As such, squirrels leap from legend and mythology to the ranks of other comic heroes. *The Unbeatable Squirrel Girl* characters and others join the ranks of squirrel cartoon and comic book characters along with Rocky the Flying Squirrel, Secret Squirrel, Princess Sally Acorn, Scrat, Skippy Squirrel, SpongeBob's friend Sandy Cheeks, and others. Some even believe that Totoro, the popular Shinto-inspired forest spirit character created by Hayao Miyazaki, is inspired by the fuzzy, gray, tufted-eared Abert's squirrel (*Sciurus aberti*) or possibly the Japanese squirrel (*Sciurus lis*). If that isn't enough to set fans aflame, former Game Freak designer Atsuko Nishida confessed that her fascination with squirrels inspired her early design for Pokémon's most famous character, Pikachu. If you have doubts, look at Pikachu's tail, which resembles a squirrel's bushy tail. Fans might also recognize the squirrel-like qualities of Emolga in the next generation, based on a flying squirrel.

In addition, if you are looking for an earworm while you are reading, check out the Kiboomers' kids' song "Gray Squirrel" or the comical "Phineas and Ferb" singalong "S.I.M.P. (Squirrels in My Pants)." You're welcome.

Besides *The Unbeatable Squirrel Girl*, many of these squirrel characters exhibit traits we often assign to actual squirrels, the same characteristics we usually attribute to ourselves, positive and negative, like having a scattered mind or being industrious. But, as we do with other wildlife species, we often get it wrong. Our casual observations might amuse us and cause us to develop a perceived understanding of the vast Sciuridae family that goes awry.

In a strange case, squirrel observation inspired a cartoon character that ultimately inspired a military term. It wasn't the squirrel's natural caching behavior at the heart of why operatives on classified missions in the US military were nicknamed Secret Squirrels. The moniker was chosen because the actual code name used by the military during the 1990 Middle East operations was Senior Surprise. Secret Squirrel, conveniently, had the same initials and could be said aloud without exposing the operation. The Secret Squirrel name was inspired by the 1965 Hanna-Barbera show *The Atom Ant/Secret Squirrel Show*. The show's segments parodied the fictional spy character James Bond, Agent 007. Secret Squirrel became known as Agent 000. Observations of natural squirrel behavior resulting in the inspiration of a cartoon character became the inspiration for a real-life military operation name.

It's remarkable to think that the world's greatest philosophers, poets, and artists have all shared with us the experience of observing squirrels, whether it was Pliny the Elder observing a native Eurasian red squirrel in ancient Italy, an ancient Chinese painter such as Qian Xuan, or John Singleton Copley painting a tiny flying squirrel tethered to a gold leash in a 1765 portrait. All of these illustrious humans were, at one time or another, amused, engaged, or frustrated enough by their own observations to include the bushy-tailed little rodents in their work. But squirrels don't discriminate in whose attention they capture. They entertain average people every day.

"Yes, I think the squirrels are great, it's so much fun to stop and watch a squirrel because you can, then, then you get stuck, then you watch until you lose sight of it, it's interesting like when it jumps and bounces, makes you happy, because it is so bouncy and happy," said one participant in a 2023 Swedish study about people and nature titled "Wildlife and Public Perceptions of Opportunities for Psychological Restoration in Local Natural Settings."

How do we perceive squirrels outside of a secret spy personification or other anthropomorphic character? We watch squirrels scurry about in parks and woodlands and attribute those movements to traits most people wouldn't want attributed to themselves. Not many of us would be pleased to be described as squirrelly, a trait the *Britannica Dictionary* defines as restless, odd, silly, or foolish. Squirrels have turned up figuratively as scatterbrained in the English language for more than five hundred years. According to the *Online Etymology Dictionary*, "The Kingis Quair" (circa 1500) contains "The lytill squerell, full of besyness," and a 1637 text includes "squirrel-headed young men." But those descriptions of perceived traits in squirrels are ultimately examples of how much we don't grasp about these creatures. Squirrel brains are more complex and squirrels more intelligent than we recognize. Perhaps being called squirrelly is actually a compliment. As in many other instances with squirrels, it may be a case of not knowing what we don't know. Despite our infinite observations throughout history and the cultural touchstones that squirrels have inspired, there is so much we don't know about this ubiquitous family of rodents. How is it possible?

*Chapter Four*

# As a Mammal of Mystery

41

*"We don't know a millionth of one percent about anything."*
—Thomas Edison

It doesn't take a genius like Thomas Edison to realize how much knowledge we lack, but it often takes humility. In the case of wildlife, frequently we know the most about the rarest wildlife and the least about the most familiar. Many squirrel species fall into the latter.

I downloaded the Squirrel Sighter app on my phone and use it regularly to record both live and roadkill squirrel sightings. It's a simple way to record the squirrels I see at home or when I travel. I've pulled over on the side of the road to document the antics of four scurrying squirrels on a nearby lawn. I've also recorded ones hit by passing cars. The app is simple to use. You open it, tap the button for a live sighting of any squirrel species, or scroll to the next screen to add data for a deceased squirrel sighting. Its simplicity creates an opportunity to feel that you are adding something to a population study, that you are adding to the bank of squirrel knowledge. Anyone using the app can be a citizen scientist or backyard naturalist. Similar to using eBird or Merlin to record bird observations, Squirrel Sighter is for everyone with a mobile phone. While the physical spotting is old-school, the data recording has moved into the twenty-first century. But the data is incomplete and nondiscriminant. Unlike eBird or Merlin, the app doesn't allow for recording species identification. In the end, there is a global map of all the squirrel sightings. While it doesn't reflect species, it does present an image of how near and far squirrel species can be found.

There are 294 squirrel species out and about in the world, so the dots on the map are numerous. Each dot represents a sighting but also reflects many squirrel species spotted across the globe. The dots beg the question: What species are they, and how are they doing? Are they all native species or expats?

Some might think that iNaturalist, an app that includes sightings and data across biodiversity species, provides a clearer picture of squirrel data. Not so, claims biologist Elizabeth Carlen with the Living Earth Collaborative, who pulled up the app to investigate squirrel sightings in St. Louis, Missouri. "According to the app, Eastern gray squirrels tended to be mostly spotted in the south part of the city," she said. "That seemed weird to me, especially because the trees, or canopy cover, tended to be pretty even across the city." She wondered what was happening with the city's squirrel population and whether the observations truly reflected the population. "Squirrels are abundant in the northern part of the city but there are no recorded observations," she posted on X, formerly known as Twitter. After taking the issue to social media with her maps, she found that others had similar problems with iNaturalist and eBird. Wildlife biologist John Vanek posted his response, "Similar pattern to what we see in Syracuse," referring to his research on melanism in the Eastern gray squirrel.

It was a case of biased data. Carlen and her coauthors published their findings in *People and Nature* in March 2024, sharing a framework for contextualizing social-ecological biases in contributory science data. They described four kinds of filters that can influence overall population knowledge. Wildlife sightings and recordings are affected by the location of the spotters—by where they have access. Another factor depends on whether people in the community are even aware of an effort to collect data and have the means to collect it. They found that often spotters don't record wildlife considered ubiquitous, nuisance, uncharismatic, or generally "boring." That can describe many undocumented squirrel sightings. If spotters aren't engaged in reporting when animals are present, it is unlikely that data will be recorded. Often, data collection is limited to when people are recreating, not while they are commuting. But even if spotters are intrigued by squirrels and would note a sighting, much of the data is collected during the day, when spotters are active but certain species, like nocturnal flying squirrels, are not, leaving gaps in the knowledge bank.

The problem with these community science apps is that the data collected influences official population reports. Those reports and the data they contain or are missing can affect land management, as city planners, environmental consultants, and others sometimes use contributory wildlife data in their work. In a way, the apps can do the opposite of what they are designed to accomplish, creating more mystery than knowledge and leading to poor management practices.

The fact that we need to learn what is happening with the populations of about 40 percent of the world's squirrel species demonstrates our significant lack of understanding about our common and unfamiliar squirrels.

Scientists across the globe are trying to increase their knowledge about squirrels. Whether with caching, migration patterns, or evolution, there is still so much to learn much about the squirrel family. The scientists have gathered together to discuss their findings at the International Colloquium on Squirrels every three years, beginning in 1994. The last they held was the Eighth Colloquium in 2018, before the COVID-19 pandemic.

Colloquium organizer Colin Lawton announced the mission for the gathering and the expansion to include additional squirrel family species in the book of abstracts for the Eighth Colloquium, held at the National University of Ireland in Galway: "The International Colloquium on Squirrels is a global event that is held every three years. It brings together squirrel researchers to discuss all aspects of squirrel biology including ecology, behavior, evolution, morphology, genetics and conservation. Originally focused on tree squirrels, the colloquium was expanded to include flying squirrels at the 4th meeting in Kerala, India, and in Galway the programme was further expanded to include the ground squirrels, so completing the whole Sciuridae family."

While that meeting shed light on what scientists knew and still needed to explore, another hasn't been held. The next one is tentatively on the calendar for 2025 in India, but details are sparse. That's just an example of how our global knowledge has met and will continue to meet roadblocks. Scarcity of funding for studies, worldwide disasters such as the COVID-19 pandemic, and wars, which make it harder to locate wildlife to study, contribute to the lack of knowledge surrounding our global squirrel species.

Squirrels aren't the only mysterious species in the world; there are many more. But those species are not seemingly as omnipresent as the squirrel. We accept the mystery of the giant squid that lives at great depths in the ocean, but the far-reaching global distribution of squirrel species makes it difficult to grasp why we don't know more. And yet it is that global distribution that can assist us in understanding why so many lack data.

Many factors make studying squirrels a challenge. Not all squirrel

As a Mammal of Mystery 45

species are as convenient to analyze as the eastern gray squirrel, which coexists in our daily lives. Many other squirrel species inhabit places that are remote. Other species, such as the elusive flying squirrels (*Pteromys volans orii*) that exclusively inhabit the small Japanese island of Hokkaido, are nocturnal, making them challenging to observe. This tiny subspecies of the Siberian flying squirrel, commonly called the Ezo flying squirrel, is adorable, with big, dark eyes and little paws, making it look like a Japanese manga character. The squirrel, although rarely seen, is so popular that its image appears on everything, even on the refillable tickets for the regional Hokkaido railway. It serves as the unofficial local mascot. But it is a challenge for researchers, who are more likely to see the squirrel's image than to see them in the wild. It took writer and photographer Tony Wu three winters to capture Japan's "daredevil" flying squirrels with his camera for the December 2022 issue of *BBC Wildlife Magazine*. If an adorable little species that is so well-known and well-loved could be that challenging to examine, imagine understanding squirrel species inhabiting more desolate areas without such a fan base. Considering a squirrel species for protection that is not as visible and easily studied is often difficult.

A team of flying squirrel researchers explored how people in Finland feel about the conservation of their protected subspecies of nocturnal Siberian flying squirrels (*Pteromys volans*) in three Finnish cities: Espoo, Jyväskylä, and Kuopio. The primary habitats for the Siberian flying squirrels in Finland are mature mixed forests dominated by Norway spruce. These woodlands, like so many others, have been changed and fragmented over the years through forest management. This ecosystem shift has led to a decline in flying squirrel populations, moving their status from nearly threatened to vulnerable. Although most Finns don't often see the small nocturnal squirrels, researchers discovered that "cities may gain an overly positive view of citizens' attitudes toward the protection of flying squirrels through current public participation methods based on self-selection procedures, such as public hearings used in land use planning."

But this is not always the case. In some areas, out of sight does mean out of mind.

While challenges exist regarding geography and species behavior in the study of many squirrel species, it can be even more difficult for scientists to investigate wildlife in areas experiencing political upheaval and war. The invasion of Ukraine in February 2022 by the Russian Federation, for example, threatened the habitat of several rodent species, including the endangered speckled ground squirrel (*Spermophilus suslicus*), the endangered European ground squirrel (*Spermophilus citellus*), and the little or pygmy ground squirrel (*Spermophilus pygmaeus*), all of which are vulnerable to habitat disruption caused by military activity.

The IUCN Red List of Threatened Species lists the European ground squirrel as endangered; however, Ukraine's Red Book lists three endangered squirrels, including the speckled ground squirrel (*Spermophilus suslicus*), Podolsk squirrel (*Sciurus podolicus*), and European ground squirrel (*Spermophilus citellus*) native to the eastern Ukrainian steppes. The war not only disturbed wildlife populations but also impacted the organizations studying them, including Rewilding Ukraine. The Danube Delta and the entire country and its wildlife were placed in doubt. Scientists in Ukraine's fields and forests were displaced by threatening land mines as the conflict continued for years, making it impossible to study and conserve the nation's wildlife amid the violence.

Severe artillery and aerial shelling between February and September 2022 put several speckled ground squirrel colonies close to the sites at serious risk. Ukrainian mammalogist Mikhail Rusin with the Kyiv Zoo, who was researching the war zone, noted that these squirrel colonies were so small that a single FAB500 bomb could quickly destroy an entire colony. All surviving colonies of ground squirrels in Ukraine are now small and fragmented, with a greater potential for threat.

"Is it unsurprising that rodents—the dominant mammal residents of steppes—faced significant population declines and became threatened with extinction. For example, in the latest list of protected species in Ukraine (often referred to as the National Red Book), 25 species of rodents are identified as threatened. And 18 species out of those 25 represent species strongly associated with Eurasian grasslands. It is no surprise that besides the obvious humanitarian crisis, the war can have a tremendous impact on steppe ecosystems, especially on protected species," alleged Rusin.

Alas, conflicts are as universal as squirrel species. They occur worldwide and adversely affect biota wherever they take place. The longer the conflict, the more damage is inflicted on wildlife, but even short bouts of violence wreak havoc.

As violence continued in Ukraine, across the ocean in Ecuador, hostility was threatening another species. Ecuador is home to Guayaquil squirrels (*Sciurus stramineus*). Described often as a robust species in size, the Guayaquil is found exclusively in the Andes of southwestern Ecuador and northwestern Peru. Aggressions in January 2024 curtailed the scientific studies of wildlife across the country, in the field and in universities. The question then becomes, How robust is the squirrel's population, and can it withstand the turmoil?

While war and other aggressions can destroy wildlife populations and conservation efforts directly, many other aspects affect biodiversity during war. Habitat destruction from toxic heavy metals, herbicides, and pollution that accompany aggressions impair a wide range of species, including squirrels. Even battle noise and resource extraction affect wildlife.

Amid all of these confrontations and trials, scientists continue the pursuit for knowledge. They wait out the wars and trudge through difficult habitats to search for small creatures and resolutions to the questions about squirrel species that persist.

"We have more questions than answers about their social and mating behavior and genetics," said biologist Karen Munroe of Baldwin Wallace University. While we know so much about squirrels, "there is still so much that's not known," according to the University of Florida's Robert McCleery, which is one reason he claims he never tires of studying them. "One of the things we are trying to understand right now is why squirrels sometimes decide to eat a seed right away and other times decide to bury it," McCleery said in 2018. Scientists wonder whether this is determined by the seed's nutritional value or the location site.

McCleery is also one of the scientists examining the decreasing population of fox squirrels (*Sciurus niger*). Fox squirrels, often described as reddish or rust colored, like one I eventually saw in Idaho, can display a variety of color phases. In the Sandhills and Coastal Plain of North Carolina, they are grayish with various patches of black on the head and feet and white patches on the nose, paws, and ear tips. Some are almost totally black with dark gray patches, while others are reddish or rust colored. Squirrels in the northwestern population typically have a tawny-brown or grizzled-gray color above, with a rusty or pale orange-brown color on the underside, ears, and legs. The top of the head is usually black, and they often have a white nose as well. McCleery hypothesizes that their population decrease may be due to competition with gray squirrels and a changing environment that favors gray squirrels over fox squirrels. He's also exploring how prescribed burns can impact populations of gray and fox squirrels in longleaf pine forests. One study found that while fox squirrels have a size advantage over gray squirrels when obtaining resources, gray squirrels become dominant as tree and shrub density increase at longer burn intervals. Such studies are another example of how information can help plan management approaches to the squirrels' habitat.

What else don't we know? According to biologist Thaddeus McRae, little study has been done on squirrel communication, but he's taking a stab at it. "The species has quite a varied repertoire of vocalizations, including a squeak similar to that of a mouse, a low-pitched noise, a chatter, and a raspy '*mehr mehr mehr*,'" he said. These vocalizations alert other squirrels, shoo predators away, and attract mates.

McRae has discovered that squirrels have "three acoustically distinct alarm calls: kuks, quaas, and moans," he said. "The most common sound by far is the kuk, which sounds like the bark of a very, very small dog. In fact, if you take a recording of a gray squirrel kuk and drop the pitch, it sounds just like a dog barking. It is usually repeated very rapidly for a few seconds before slowing down. . . .

"Kuks and quaas are used to scare off predators and to warn other squirrels. Both are noisy, scratchy sounds with no clear note or tone," McRae said. "In contrast, the vocalization known as a moan has a very clear tone, which usually quickly rises and slowly falls, sounding very like a sad person moaning. Quaas and moans sometimes blend into each other, and sometimes that makes them sound remarkably like a disgruntled chicken. Squirrels often use more than one call, but rapid kuks and quaas are most often given when a terrestrial predator is around, and moans are usually used in response to aerial threats like hawks."

Even with all that McRae has discovered, his research raises more questions, including whether there are regional variations in squirrel chatter, as with the vocalizations of some songbirds and wolf howls.

Mammalian ecologist Stan Boutin of the University of Alberta in Edmonton has been studying the behavior of red squirrels in boreal forests in the Yukon for over three decades. He's climbed more than five thousand spruce trees and prides himself as having fallen just once. He's marked squirrel pups with ear tags in a square kilometer (about 247 acres) of boreal forest between Yukon's Haines Junction and Kluane Lake. He's discovered that the tiny reds spend their entire lives in a small forest patch surrounding a midden, their pile of spruce cones complete with tunnels and chambers. While squirrels have one litter of pups in an average year, he's uncovered that squirrels have two litters and sometimes even breed into the last days of summer when spruce cones are abundant during mast years. "These buggers are cranking out reproduction when it's ahead of masting reproduction."

However, with all of his squirrel observations and discoveries in the northern boreal forests, Boutin has more questions. How can female squirrels predict a good food supply before it exists? Is there some chemical signal within the spruce tree buds that triggers what Boutin calls "adaptive anticipation" within these tiny orange bushy-tails' boreal forests?

He wonders what's up with the fluorescence of flying squirrel fur that makes them look like they should be on a Peter Max poster. A 2022 study investigated the neon pink that black light picks up in squirrel fur. Some researchers hypothesize that the strange colors on the squirrels' bellies might help camouflage them from predators such as owls that can detect ultraviolet light, causing the predators to mistake the squirrels for lichens and other glowing plants. Another hypothesis is that the fluorescence could be a trait that assists squirrels in mating or communication.

Researchers in Italy are delving deeper into the population of European red squirrels threatened by infectious diseases, habitat fragmentation, and their invasive gray cousins. They have uncovered a case of lymphoma that led to an ectopic pregnancy in a Eurasian red squirrel. Because wildlife can be a "sentinel for pathologies that have environmental causes and that could affect other animals and humans" in the same environment, exploring cancer incidence in wildlife, including squirrels, is crucial. Little is known about Eurasian red squirrel reproductive pathology. The researchers concluded that further studies are needed to "clarify the environmental drivers" of cancer in wild squirrel populations and suggested the development of a wildlife cancer registry.

Other squirrel research has delved into how the rodents regulate their blood chemistry by removing glucose and electrolytes such as sodium and storing them elsewhere. This exploration could help explain how other hibernating animals stay hydrated and might lead to discoveries that help humans deal with conditions such as diabetes.

While the research is promising, questions persist. As scientists strive to fill in the information gaps, will there be enough funding and opportunity to conserve the world's squirrel species and assist us in understanding their place within our natural world in this period of human-induced climate change? In the meantime, humans continue their long history of observing wild squirrels in woodlands and parks.

*Chapter Five*

# As an Ambassador

*"Ambassadors are by definition foreign bodies."*
—Clive Sinclair

While I was observing a native American red squirrel on that beautiful autumn day seaside in Maine, in city parks across the country, including New York, Boston, and Philadelphia, residents and tourists were becoming equally mesmerized by the antics of local resident urban squirrels. For millions of people, city parks are the place to view squirrels and, for some, for the very first time. For others, visiting urban parks to see squirrels is a beloved tradition that many repeat regularly.

It would appear that squirrels and city parks have been in eternal synchronicity. But, surprisingly, urban squirrels are relatively new residents in our urban wildlife landscapes. Of course, long ago, before humans developed urban centers, squirrels ran wild everywhere, including on the land now beneath our cities. However, America's growing European population had pushed native squirrels out of developing cities by the mid-nineteenth century.

Squirrels found their way back into these small green spaces and large urban parks beginning in one of America's first cities, Philadelphia, Pennsylvania, the City of Brotherly Love. Before European colonizers founded the city of Philadelphia in 1682, the Lenape people inhabited the land. They lived on and ate from the abundant landscape, hunting and fishing for many animal species, including deer, bears, beavers, and squirrels, which they called *xanikw*.

The story of the Lenape's ancient relationship with squirrels was shared by tribal member Nora Thompson Dean in 1977. The story is recorded on the *Lenape Talking Dictionary* website. "Këkhìtkil Hùnt Na Xanikw: Squirrels Were Said to Be Huge" explains how Squirrel, once a huge man-eating animal, became a tiny creature. It is said that Squirrel ate everything he caught: "He ate various animals, weasels and skunks and lynxes, and all the animals when he caught them, he ate them. Finally all at once a two-legged person was running along, he must have been a Lenape." Squirrel caught the person, tore him to pieces, and ate, an action angering the Creator. "You have truly done something strange! You have killed my child!" ("Kehëla yukwe kchipaihosi! Knilao nichan!"). Squirrel was punished for his behavior by being transformed into the tiny creature we know today. It is believed that when that change occurred, the relationship between humans and squirrels shifted and squirrels became a food source for humans.

Jeremy Johnson, cultural education director for the Delaware Tribe of Indians, shared this story with me and other visitors at Pennsylvania's Hawk Mountain Sanctuary on an April day. He added that the age of that tribal story coincidentally corresponds with the extinction of the massive, ten-foot-tall giant short-faced bear (*Arctodus simus*) that inhabited North America during the Pleistocene epoch, some eleven thousand years ago. While the Lenape hunted squirrels, their relationship with the small rodents encouraged the species' survival. As Europeans flowed into the New World, however, the Lenape and much of the wildlife were displaced by the growing population and development. William Penn and Quaker colonizers created the colony of Pennsylvania with a 1683 treaty signed at what is now Penn Treaty Park between the colonists and the Lenape. The Lenape were eventually forced to relocate to reservations, taking their knowledge of the land and creatures with them.

Penn envisioned the city of Philadelphia as not unlike the rural English towns he knew. There would be gardens and orchards for his "greene country towne." While he envisioned private lots of at least a half acre, each with a house in the middle, Penn also set aside public land to form five public squares with multiple uses, including militia drills and burial grounds—public green spaces without a plan for wildlife.

Explorers in this pre-Darwinian, pre-Linnaean early American period recorded their observations of the wild flora and fauna they encountered on trips throughout the rural New World landscape while the cities were growing. One of the earliest to describe squirrels was John Brickell, a native of Ireland, who traveled to the mid-Atlantic region and recorded his observations of four sorts of squirrels in his eighteenth-century book *The Natural History of North Carolina*. His observations included the fox squirrel "being the largest and smelling like a Fox," "larger than a Rabbet," having "good Meat," but "very distructive and pernicious in Corn Fields." Aside from his physical description of gray squirrels, he noted that the "Fat of these Squirrels is Emolient, and good against Pains in the Ears, and the Teeth, are said to be used by Magicians in foretelling things to come." His description of flying squirrels included their behavior as "being unable to bear the cold and severity of the weather, and generally half a dozen or more lie together in one nest, which is always in a hollow tree, and have their stores of provisions near them, whereon they feed during cold weather," but he added that while they are "easily made tame," they are "Enemies to Cornfields (as all the other Squirrels are) and only eat the germinating Eye or Bud of the Grain, which is very sweet." Although his ground squirrel observations included that the rodents had "much the same Virtues and Uses with the other sorts of Squirrels," they "may be kept tame in a little Box with Cotton in it."

His observations of squirrels as farmland pests rang true in Pennsylvania as well. The Anglo-American settlers used existing Lenape farmland and clear-cut swaths of forests to build their villages and towns, including Philadelphia. This sent squirrels from the woodlands into the growing farmland outside the city to forage on corn and wheat crops, where they were hunted as pests and used for food and fur. Tens of millions of gray squirrels were killed in America during the seventeenth and eighteenth centuries.

Squirrel-hunting accounts were frequently chronicled in early American newspapers, including a 1791 report in *The Gazette of the United-States*, "A correspondent from Johnston informs, That last week, two small parties, with their guns, went out a *Bird-Hunting*—and from Monday till Thursday killed as follows:—One party 5030 birds and 690 squirrels—the other, 4228 birds and 670 squirrels."

Brickell and others recognized squirrels as agricultural pests. In the surrounding Pennsylvania countryside, where many residents had moved from, the native squirrels were plentiful and continued to be viewed as agricultural pests. In 1749, Pennsylvania's Game Commission paid bounties on 640,000 squirrels.

Early Philadelphia naturalist John Godman remarked on those bounties and the squirrels' pestilential persistence that triumphed over humans' laws and money: "This species . . . was once so excessively multiplied as to be a scourge to the inhabitants, not only consuming their grain, but exhausting the public treasury by the amount of premiums for their destruction. 'Pennsylvania (says Pennant) paid from January, 1749, to January, 1750, *eight thousand pounds* currency; but on complaint being made by the deputies that their treasuries were exhausted by these rewards, they were reduced to one half.' . . . How improved must the state of the Americans then be, in thirty-five years to wage an expensive and successful war against its parent country, which before could not bear the charges of clearing the provinces from the ravages of these insignificant animals!"

But Godman did not limit his declarations to squirrel bounties. He also noted his admiration for gray squirrels: "This species is remarkable among all our squirrels for its beauty and activity. It is in captivity remarkably playful and mischievous, and is more frequently kept as a pet than any other. It becomes very tame, and may be allowed to spend a great deal of the time entirely at liberty, where there is nothing exposed that can be injured by its teeth, which it is sure to try upon every article of furniture, &c. in its vicinity."

Squirrels had two distinct roles in American culture: pets and pests. Godman mirrored the feelings of many in Pennsylvania and beyond who enjoyed keeping squirrels as pets, including Benjamin Franklin. While squirrels had always been kept as pets, Franklin's pet squirrel was one of the few squirrels in Philadelphia at that time; it would influence the pet trade in Philadelphia and other cities in the future.

Franklin wouldn't be the last politician to champion the practice of keeping squirrels as pets. As seen in Brickell's writing, having pet squirrels was common. Other politicians besides Franklin included, years later, President Warren Harding and his wife, Florence. They kept a pet squirrel named Pete in their White House family. Pete even showed up at White House press conferences. President Harry Truman also had a pet squirrel named Pete. In addition, Theodore Roosevelt had a pet flying squirrel at his family's Sagamore Hill home.

Pets and pests, roles that continue. But another role for squirrels was on the horizon, the role of ambassador. And that would begin in the growing city of Philadelphia.

Franklin's Philadelphia friend, famous naturalist and explorer John

Bartram, traveled with his son, William Bartram, to parts of Carolina and Florida. His son wrote about several tree, ground, and flying squirrel species found on these famous explorations through the southern British colonies in North America from 1773 to 1777. He noted that the "ground squirrel, or little striped squirrel of Pennsylvania and the northern regions, is never seen here, and rarely in the mountains northwest of these territories." The Bartrams' riverside home in Philadelphia became the young country's first botanical garden. Gray squirrels still inhabit the historic property, so much so that the homestead's gardeners must place mesh down in the bulb gardens to prevent the squirrels from interfering with the plantings.

However, gray squirrels were not found in the young city's parks in those early years. Not everyone shared William Penn's view of greening the young city. While squirrels remained in the fields outside the burgeoning urban center, Philadelphia grew more and more crowded and dirty. Poor immigrants from Germany, Scotland, and Ireland poured into the Port of Philadelphia, with more than 150,000 arriving by the mid-1700s. The streets became crowded as industry developed. Wildlife took a back seat to the city's congested path.

Like other new American cities, Philadelphia suffered from its growth in the eighteenth century. Roads were unpaved and littered with garbage. The city's squares were not used for rest or relaxation. They were used for clay mining, trash dumping, and public executions. But despite the filth, Philadelphia grew into a major city with schools, libraries, theaters, and newspapers—and prominent residents such as politician, polymath, and squirrel aficionado Benjamin Franklin.

After the American Revolution, Philadelphia's population surged, and by the 1840s, the city suffered from even more overcrowding and disease where row houses and tenements flourished. Violence filled the City of Brotherly Love. The wild squirrels that once inhabited the land were long gone but not forgotten.

"This animal is so well known as to need no further comment, except to call attention to its differences of color and size from the northern race," stated the description of the Carolina gray squirrel in *The Mammals of Pennsylvania and New Jersey*. To city residents, however, squirrels were far from being thought of as common pests and instead had become known as exotic creatures of the deep woods and a memory from country living.

"Squirrels might seem too small and commonplace to have been historical actors, but they are on the stage frequently in the American story," wrote Dan Flores in *Wild New World*. He couldn't have been more correct. They were about to step right onto the stage during this turbulent time in Philadelphia, years after the death of Benjamin Franklin. In response to the city's woes, an idea sprang up to reintroduce squirrels from the countryside to Philadelphia's North East Publick Square, which had been renamed Franklin Square.

Today, the square bordering the historic district is filled with children riding on the Parx Liberty Carousel and playing miniature golf. It barely resembles its history as one of the five original open-space parks Penn had planned. It already featured a garden and a fountain. Now, that same fountain has choreographed light shows. The park's rich history as a burial ground, cattle pasture, and magazine storage for ammunition powder before becoming one of Philadelphia's parks has become a distant footnote. The street bordering it, Sassafras Street, had so many horse races that it was eventually renamed Race Street. The city square turned park would soon become the first to introduce squirrels into the urban landscape and launch a new role for American squirrel species.

Named after squirrel pet owner and founding father Benjamin Franklin, the park was the perfect spot for this social experiment. In May 1847, three squirrels, brought from the surrounding countryside or someone's tenement, were set free in the square with food and shelter boxes.

"A correspondent calls our attention to the fact that someone has placed in the Franklin Square three gray Squirrels, and those who visit it early in the morning may have an opportunity of seeing them skipping about from branch to branch, with an activity peculiar only to themselves. Two boxes have been placed in one of the trees, one containing 'nuts for them to crack,' while the other answers for retreat at night," reported the *Philadelphia Ledger* on May 11, 1847, under the heading "Local Affairs." The news fell between news of Tom Thumb and an enormous steam boiler. This arrangement had already hooked the *Ledger*'s correspondent, who continued, "But this arrangement does not seem to suit one of the most active; it has commenced making a home for itself in one of the large maples near the western gate. It is earnestly to be hoped that persons will not molest these harmless creatures, which add much to the beauty of the place."

Regardless of where they came from, those first squirrels must have been surprised at their new surroundings in Franklin Square. Just like the one gray that had already moved into the tree at the gate, others soon spread to the surrounding streets. The population was allowed to thrive despite concern from residents who opposed the introduction, fearing the squirrels would prey on the city's songbirds and nests. By the mid-1800s, squirrels figured in the picture of the ideal city square.

"With fountains glancing, birds singing, children playing, squirrels chasing each other from bough to bough, peacocks strutting in all the glory of their rainbow plumes, sportive fawns skipping to and fro, and clambering vines and other shrubbery varying the monotony of trees and grass, these squares would be truly delightful resorts, affording the means of increasing enjoyment to the increasing multitudes that throng this metropolis."

But in 1864, those voices of opposition weren't ignored, even though there wasn't any proof that the squirrels had disrupted the songbirds. The nest boxes were removed, and the squirrels were captured or killed. But the legacy of those squirrels continued.

On a rainy Philadelphia day, science historian Etienne Benson sat with me in the library of the University of Pennsylvania, where campus squirrels have had their own social media account since 2016 and Benson is an adjunct professor. Youthful-looking Professor Benson sat in his raincoat looking at me through his dark-rimmed glasses, discussing the history of eastern gray squirrels brought to urban parks first in Philadelphia, just blocks away from where we sat, then to Boston, Massachusetts; New Haven, Connecticut; and, finally, New York City.

The common eastern gray squirrels at the center of these parks were given their official scientific name, *Sciurus carolinensis* ("squirrel of Carolina") in 1788 by German naturalist Johann Friedrich Gmelin (1748–1804).

Benson's research focuses on the history of science, politics, animal–human relationships, and the environment. His definitive paper, "The Urbanization of the Eastern Gray Squirrel in the United States," was published in the December 2013 issue of *The Journal of American History*. It begins by noting a July 12, 1934, radio talk given by the recently retired chief field naturalist for the US Bureau of Biological Survey, Vernon Bailey, titled "Animals Worth Knowing Around the Capitol." Bailey singled out the common gray squirrels as "probably our best-known and most loved native wild animals, as they are not very wild and, being very intelligent, accept and appreciate our hospitality and friendship."

Bailey wasn't the only person enjoying a population of urban squirrels by then. Some might have thought it was the culture, finance, or influential people that impressed Europeans visiting our cities. While all of that played a part, it was perhaps the cordial relationship between New Yorkers and Central Park squirrels that made the greatest impression on visitors.

Philadelphia was just the start for eastern gray squirrel introductions into parks from nearby rural woodlands between the 1840s and the 1860s across the country as urban park projects took shape—to Boston Common in Massachusetts in 1855, to New Haven, Connecticut, and, finally, to New York City's Central Park, where a few gray squirrels were released into the thirty-eight-acre woodland known as The Ramble by the Central Park Menagerie. And so began the admiration of squirrels and their distribution to American urban parks.

"A wide variety of institutions and individuals," wrote Professor Benson in his hefty, illuminating article, "shaped the moral-ecological character of urban public spaces. . . . Even the least powerful members of human society could demonstrate the virtue of charity and display their own moral worth."

The quintessential squirrels of New York City's Central Park embarked on a life that influenced their history but also the history of the residents of New York and beyond. That's a broad statement, but if it weren't for the generous treatment given to Central Park's squirrels, our country wouldn't have had the standing it gained on the world stage. Ernest Thompson Seton encouraged the feeding of park squirrels as a way to "cure boys of their tendency toward cruelty."

While fashionable New York City ladies sported squirrel fur-collared coats, the kindness toward squirrels that ensued in the park became synonymous with lessons of kindness, compassion, and character formation. The New Yorkers' attitude toward squirrels was used to measure the level of civilization among Americans, which was noted by visitors from near and far.

American naturalist John Burroughs, writing from his home in the wilds of the Hudson Valley, brought his local wildlife to his readers and inspired land and wildlife preservation. He has been credited with creating the modern nature essay, which encouraged his readers to enjoy the wildlife around them, including squirrels. In his 1875 book *Squirrels and Other Fur Bearers*, Burroughs wrote that the squirrel was an "elegant creature" that "excites feelings of admiration akin to those awakened by the birds and the fairer forms of nature." The Victorian belief of humans living in harmony with nature that Burroughs, Walt Whitman, Henry David Thoreau, and others spread through their writing heightened interest in Central Park's resident squirrels.

But squirrels' popularity and the population numbers of the rodents would wax and wane through the years, and by 1883, the population had grown to a point that the city organized a series of early-morning squirrel hunts, despite opposition from the newly formed American Society for the Prevention of Cruelty to Animals (ASPCA), to control the thriving population. After one of the hunts, the Central Park Menagerie's director reported that "fully a hundred fine fat squirrels" were killed. Ultimately, those hunts became illegal, and the population was allowed to continue.

While the population grew, so did its media presence. During this time, there wasn't a lack of squirrel stories in the New York newspapers, many in *The New York Sun*. Amid reports of missing corpses and recent weather reports, stories of squirrel sightings entertained and informed city residents and visitors.

*The Sun* reported on a sighting of squirrels in Central Park in 1888: "A round dozen, by actual count, were seen playing together on the grass of a well-shaded hollow in the Park the other morning. They indulged in the squirrels' favorite little game, alternately standing erect and jumping over one another's backs."

But debates continued about whether the squirrels contributed to society, how they should be treated, and whether they were indeed pests. In the 1890s, stories of squirrel control measures, accusations of Italian laborers hunting squirrels for food, and numerous people feeding the colony regularly appeared.

In 1900, *The New York Sun* reported on the public's admiration for Central Park's squirrels in an article aptly titled "Endless Source of Delight to Visitors, Young and Old." The article estimated the number of squirrels in the park at that time to be around 150, although many thought their numbers were quite a bit higher. Numbers inevitably varied from year to year. It was noted that the park department even practiced removing some of the Central Park squirrels to spark squirrel populations in other city parks, which spread the famous squirrel love to many New York City borough residents.

In the short time since their release into the park, the squirrels inspired a fan club of visitors who fed them regularly. As time passed and the squirrels became accustomed to hand-feeding, they grew tamer and tamer. That same 1900 article in *The Sun* included stories of the individuals who were visiting the squirrels and their regular encounters with squirrels that were becoming wild pets: "A lady, driving, halts at some favorable spot, where perhaps she has stopped before, and taps with a nut gently on the rim of one of the carriage wheels. A squirrel comes near, his bushy little banner waving; coming nearer with his characteristic approaches, he makes finally a dash for the carriage and flies up the wheel to be rewarded with the nut at its summit."

Not unlike the visitors to Central Park today, who use cell phones to snap selfies with squirrels, many carried film cameras in 1900. "Here for example was a small boy walking along a path with an older sister, small boy bright and earnest and carrying a camera. Sister with a camera, too. They stop to look at a squirrel running along on the grass near the path and the squirrel stops to look at them and sits up and regards them, the small boy in particular, with a sort of puzzled wonder. But the small boy has been focusing him all this time, and the instant the squirrel settles in that attitude, snap! goes the boy's camera, and the squirrel jumps, and snap! his sister has taken him as he jumps. Now will they, or will they not, these two young people, enjoy looking at those two pictures when they get them finished. An endless source of amusement and delight are the squirrels of Central Park."

Everyone, rich and poor, residents and visitors, enjoyed Central Park's squirrels. Even the park commissioner, William R. Willcox, got involved in the well-being of the park's red and gray squirrels by ordering their feeding during the harsh winter of 1901. The *New-York Daily Tribune* highlighted his initiative in "Willcox Squirrels' Friend: The New Commissioner Provides Peanuts for Hungry Park Dwellers" on January 11, 1902. Commissioner Willcox "discovered that the children's summer friends were not fed when the snow was on the ground" and "ordered that they be fed."

But not knowing much about squirrel behavior and biology, one of the park's directors, John Smith, delivered nutritionally poor raw peanuts to the snow, believing that heavier native hickory nuts would fall beneath the snow and be more difficult for the squirrels to find than the lighter peanuts. Fortunately, the park squirrels survived the error. Their care extended over the years, and they were also supplied housing by the enthusiastic, squirrel-loving city community.

Squirrel houses weren't a new invention. A drawing of a squirrel house complete with a tethered red squirrel sitting beside it can be found in an illuminated manuscript, the *Book of Hours of Catharina van Wassenaer*, written and illuminated in the Netherlands around 1490 and housed in the collection of the Morgan Library & Museum in New York City. Undoubtedly, there were many others. But the history of squirrel houses was about to launch into a new phase, one of elegance befitting their park community sandwiched between the Upper West Side and Upper East Side of Manhattan.

By 1907, Central Park's squirrel colony was so well-loved that it received "the grandest" of New Year's gifts, "none other than a superb rustic apartment house, designed to accommodate three families and equipped with a veranda as hospitably broad as that of a seashore hotel, a roof garden modeled after that of the Presbyterian Hospital and an open-air gymnasium that vies with any operated by the Borough of Manhattan," from society donors Mrs. Russell Sage and General and Mrs. Charles Dodge.

Had these socialites had the means in 1907, their over-the-top gift to Central Park's squirrel colony might have reached the heights of the elaborate squirrel home that Derrick Downey Jr. showcases on his 2024 TikTok account. His local Los Angeles, California, neighborhood squirrels, Maxine and Richard, enjoy a house with its own Ring doorbell camera, furniture, Christmas tree, and working TV.

While the interactions with Central Park squirrels instilled admiration for nature and the kind treatment of wildlife, feeding park squirrels was also seen as an analogue for the responsibility to the poor during this early period in America. The author of the 1906 article "Squirrels and Men" in *The Evening World's Daily Magazine* wrote, "In one respect squirrels do not differ from men. Gratuitous charity is as demoralizing to one as the other." Marian Longfellow published a poem in 1908, "The Pensioner in Gray," in which the hand-fed gray squirrel symbolized the dependent elderly.

As the squirrels' popularity grew, so did their population, but while Central Park's squirrels were getting food and housing, most of the squirrels introduced to some of America's other parks faced a different fate. There needed to be enough natural tree nuts for the squirrels to eat or enough habitat for adequate shelter. Squirrel populations that weren't fed or given winter nest boxes didn't survive. This was the case for the squirrels released in Philadelphia's Franklin Square, the New Haven Green, and Boston Common.

Not so in Minneapolis, Minnesota. In the early 1900s, Theodore Wirth, the superintendent of Minneapolis parks, declared war on Loring Park's native American red squirrels. The predominant squirrels in the parks were seen as vicious predators that ate baby songbirds and protected their own territories. When he heard about the introduction of friendly gray squirrels, known affectionately as "cream puff" squirrels, into Central Park, he sent his rangers to execute the red squirrels living in Loring Park. He then shipped about two dozen gray squirrels from Washington, DC, to supplant the red population, even providing them with nest boxes. "Mr. Wirth has confidence they will appreciate the small sized cottages in the trees he has had arranged for their comfort and that they will not run away," reported the *Minnesota Star Tribune* on May 20, 1909. Wirth declared the mission a success in his 1919 annual report. Today, a healthy population of gray squirrels still exists.

Many of those first introduced park populations did perish from lack of food and habitat, but also from the opposition of bird-loving residents who felt the new residents threatened the wild birds. This downturn was temporary, as Frederick Law Olmsted's park plans extending to other parks outside of Central Park provided urban green spaces that were much more habitable for squirrel populations than the earlier public squares, commons, and city greens. Those sweeping urban green spaces with woodlands were perfect for squirrels to flourish in.

By 1920, the squirrel colony in New York's Central Park was booming. Their numbers had grown to an estimated five thousand. But another waning period was on the horizon, as a nascent ecological movement was rising in the popularity of squirrel predators, such as hawks and falcons, bringing about a decline in the city's squirrel population. Feeding the wild rodents became less a boon of character and more a break of ecological values.

But this downturn didn't curtail America's fascination with the adorable bushy-tailed rodents in and out of city parks and wild spaces. In the coming decades, squirrels will continue to entertain Americans in parks and on theater stages. Tommy Tucker, a trained male eastern gray squirrel, was dressed in women's clothing and taught to perform tricks in the United States from 1942 to 1949. The squirrel celebrity was even employed to sell US war bonds during World War II. A *Washington Post* columnist called him "the most famous squirrel ever to come from Washington."

Urban parks were just a part of greening the American urban landscape. As green parks were added to America's cities, the same newspaper writings and the writings of Thoreau, Ralph Waldo Emerson, Burroughs, Roosevelt, and others inspired the preservation of wild spaces across the country in the form of national parks. And with those parks came the resident welcoming committee of native squirrels, which didn't need to be introduced into the habitat; they were already there.

While squirrels continued delighting visitors in urban parks, they also served as ambassadors to the throngs of visitors to many of America's national parks. And they have continued to do so. One of the little park squirrels is the Uinta ground squirrel (*Urocitellus armatus*, formerly *Spermophilus armatus*), a unique and vital part of the Greater Yellowstone Ecosystem that is often overlooked amid Yellowstone National Park's megafauna. These grayish squirrels are essential prey for long-tailed weasels, hawks, badgers, coyotes, and grizzly bears. In fact, the *Jackson Hole News & Guide* referred to the squirrel as the "salmon of the interior West." Although they inhabit grasslands, sagebrush, and mountain meadows up to elevations of eleven thousand feet, these squirrels duck back under the ground to hibernate for eight months beginning as early as July.

Perhaps one of the best-known park squirrels is a cousin of the Uinta, the large, seventeen- to twenty-one-inch-long rock squirrels (*Otospermophilus variegatus*, formerly *Spermophilus variegatus*) that inhabit one of the country's most visited national parks, Grand Canyon National Park, with its nearly five million annual visitors. Those rock squirrels, however, don't have the same reputation as the friendly eastern gray squirrels living the high life in New York Central Park's Ramble. Called "unforgiving" and "absolutely ferocious" by park rangers, these wild squirrels are considered the most dangerous animal in Grand Canyon National Park. The hungry, tourist-fed rock squirrels can also be found in Zion National Park.

Consider the warning post on Grand Canyon National Park's Facebook page on May 24, 2021, that included a photo of a squirrel in front of the Keep Wildlife Wild trail sign. "Enjoy squirrels from a safe Distance. Their sharp teeth crack nuts—and cut fingers. On a busy holiday weekend, as many as 30 park visitors with bleeding fingers will require medical attention. If squirrels approach you, back away and give them space." One park visitor, Vickie Olson, shared the story online of how one of the squirrels jumped into her lunch box and "grabbed the whole entire apple and ran off. Greedy little buggers."

Whole apples aren't in the squirrels' natural park diet, and *greedy* is perhaps less accurate than opportunistic, but the situation can't be ignored. In the wild, those squirrels feed on pine nuts, cactuses, sumacs, grasshoppers, and even animal carcasses, but trouble begins when tourists feed them. The squirrels are naturally not aggressive, but they have become biters with increased human encounters. While feeding them is illegal, not all visitors obey that rule, and some, like Olson, suffer from the bad behavior of others who don't follow the rules. Feeding the squirrels removes their natural fear of humans and triggers them to search out the accessible snacks they can obtain from them. Those human-acquired snacks are also dangerous to the health of the squirrels because they don't have the nutritional value of their natural food. They get full bellies without the necessary nutrients and minerals to survive in the wild.

Although rock squirrels may dominate the park's squirrel stories, Grand Canyon National Park is home to other squirrel species, including Abert's squirrel, the rare Kaibab squirrel, the golden-mantled squirrel, and the spotted squirrel, among others. Those uncelebrated, nonaggressive squirrel species will most likely fade into the photos visitors take of the wide canyon vistas. I wonder how many tourists might find one of these peering out from a tree if they enlarge their photos.

Those squirrels, however, do have their own interesting story. The Kaibab squirrel of the park's North Rim and Abert's squirrel to the south, both tassel-eared species, evolved from the separation caused by the physical barrier of the Grand Canyon about ten thousand years ago. Visitors will see that while they both retained their ear tufts from their common ancestor, they developed their own distinctive coloration in separate areas of the canyon.

But let's be honest: Visitors to the Grand Canyon aren't visiting the majestic park for the squirrels. Like many national park nerds, I had my sights set on getting my National Park Passport book stamped and heading out on the trail leading to the canyon's rim to see the popular expanse on my visit in 2019, but that changed soon after I stepped out of our car and saw one of the park's rock squirrels right outside the visitor center. That large squirrel, appearing like a boisterous park welcoming committee member, stopped me in my tracks. And then I saw others. Their large size and their behavior were as unexpected as the sight of the large spectacled flying foxes that took my breath away during a visit to Queensland, Australia, in 2016. These squirrels and their size demanded my attention and photographs. But despite all this attention, they were still squirrels: ubiquitous at the South Rim but, for many, serving solely as background wildlife to a dramatic canyon setting and a collection of megafauna, including bighorn sheep, that seize the attention of most visitors.

Ultimately, those national park squirrels indicate a complex relationship that continues to develop in America with our charismatic, omnipresent squirrel species. As welcoming as those rock squirrels are in our national parks, the Facebook-posted comments about their behavior demonstrate the spectrum of feelings all squirrels evoke in humans, from pure joy to raging irritation. History is filled with the waxing and waning of population numbers of both native and nonnative species, along with the waxing and waning of our attention in and out of our parks, where squirrels, native and nonnative, show up to greet visitors and residents alike.

## Chapter Six

# As a People Greeter

"FEEDING THE SQUIRRELS IN UNION SQUARE."
From the Painting by W. ST. John Harper
1882

*"A Walmart greeter is an employee whose role is to wait at the front door of a Walmart and greet all shoppers who enter."*
—Sarah Nassauer

The introduction of squirrels into urban parks spread across the United States from east to west. Julia Davis Park in Boise, Idaho, received its introduced squirrel population in the early twentieth century. Unlike many of the parks on the East Coast experiencing gray squirrel introductions, Boise's park was the recipient of fox squirrels (*Sciurus niger*).

Julia Davis Park is the city's oldest park, created in 1907. It rests near familiar American food chains, including Trader Joe's and Chick-fil-A, which provide touchstones for visitors. The historic forty-acre park houses museums such as the Idaho State Historical Museum, Pioneer Village, and the Idaho Black History Museum. A rose garden with trees is just a short walk across from access to the well-visited Boise River Greenbelt, with its tree-lined paths alongside the Boise River, in view of the state university. During my visit to the park and its museums, I spied a descendent of those early fox squirrel introductions.

The fox squirrel I spotted scampered around the rose garden and up a tree. As a squirrel species, its body shape and bushy tail were as familiar as the food outlets nearby and the gray and red squirrels at my home in New York's Hudson Valley, but this species of tree squirrel had a yellowish belly and was larger than both of them. I wasn't aware at the time that the squirrel wasn't a native species, finding out only later that it was introduced as a result of the wave of positive introductions in other cities.

Julia Davis Park's website doesn't mention its squirrel population. However, it does highlight birding as a park amenity, a popular outdoor activity for wildlife enthusiasts. Not many visitors venture out to specifically spot squirrels. But for me, the accessible squirrels were as much an amenity as the birding trail and greenbelt access the park offered. Just like Walmart greeters, their friendly presence became my first welcome to the Pacific Northwest state.

Like the small red squirrel I observed in Maine, this Boise fox squirrel held my attention as I photographed it climbing up and down the park's trees. I was as eager as those two children in Central Park were to watch the squirrels over a century ago and the squirrel-watching tourists historian Etienne Benson later shared with me that he observed in Boston's Harvard Yard, which led him to study their historic introductions. Like others who see squirrels on their visits, this squirrel allowed me to engage with somewhat familiar nature in the middle of an unfamiliar city.

Perhaps it's that friendly squirrel welcome that so many college students, experiencing life away from home for the first time, find when they move onto college campuses across America. Squirrels have become so beloved on college campuses, regardless of species, that Jonathan Gottshall, a squirrel lover and California resident, set up a website that ranks campuses on their squirrel populations, including the little critters' appearance and personality. Prospective students can check out the university's campus squirrel population along with dining options and dorms. The ratings are based on submissions by college students, employees, and others. And if a healthy, friendly squirrel population is a top criterion, becoming a Fighting Squirrel at Mary Baldwin University in Virginia might be an aspiration. The Fighting Squirrels' team nickname was inspired by the squirrel on the family crest of the university's founder, Mary Julia Baldwin. Gladys the Squirrel, the mascot from the 1980s, serves today as an ambassador to the alums of Mary Baldwin College for Women and is seen at homecoming and other events. But that family crest squirrel originated in Ireland, home to native Eurasian red squirrels, a cousin to the American red squirrels found in Virginia. However, students on campus are more likely to see an eastern gray squirrel or a fox squirrel daily than a native red.

*The Purdue Exponent*, published by Purdue University students, features a Squirrel of the Week feature to highlight squirrels' antics on and off campus. And students at the University of Pennsylvania can follow @upennsquirrels on Instagram to check out the squirrels inhabiting their Philadelphia campus.

When I sat down with science historian Etienne Benson the previous spring at the gray squirrel–friendly Philadelphia campus of the University of Pennsylvania, he shared the memory of seeing foreign visitors taking photos of the gray squirrels on his Boston Harvard campus. Those squirrels were as exotic to the visitors as a Eurasian red squirrel would be in Philadelphia or Boston.

Suppose a student wants to do more than enjoy the antics of squirrels scurrying across their college campus. In that case, they can head to the Warner College of Natural Resources at Colorado State University for Fat Squirrel Week, when students and @csusquirrel_ submit photos to the college's Instagram account over a ten-month period, culminating in a vote and Squirrel Sunday celebration.

To get further into squirrel studies, the University of New England in Maine has a unique research project that has gathered data since 2010. The university's Project Squirrel involves collecting data by catching, measuring, ear-tagging, and radio-collaring resident campus gray squirrels.

I traveled to Maine again one sunny October day to speak with Noah Perlut and his students about their Project Squirrel research. The Biddeford, Maine, campus, where they study and work, rests alongside more than four thousand feet of picturesque shoreline, an area used as a corridor for wildlife, including squirrels. The school even has its own island serving as a living laboratory in Saco Bay and an oak forest a short walk away from their classroom. It's a perfect place to observe squirrel populations during the busy autumn season, when the rodents are actively preparing for winter.

The university has its roots within conservation-minded Franciscan monks on the banks, and stewardship of our planet is baked into the school's mission. This justifies its place to nurture the next generation of squirrel researchers.

Perlut noted, "When I first started the project I looked at the literature and couldn't believe there was not a single published ecological study on gray squirrels in New England." There wasn't information on basic behaviors, such as diet, breeding season, body size, or home range, he claimed. "Most basic info about the most common animals that you see was unknown."

Perlut explained the squirrel studies as the guest professor in a introductory ecology class; he introduced the students to Project Squirrel and the field biology techniques the practicum would include. A brief discussion ensued about the many species of squirrels on campus, including red squirrels, chipmunks, flying squirrels, and groundhogs. Even when people see squirrels every day, they don't see the full picture. Perlut explained again the reason for beginning the popular gray squirrel project on campus. "Nobody knew anything about them. How long they lived. How much space they needed, their home range. What are they actually eating? Do all these factors matter based on their sex? Or the season?" The surprise in the eyes of these students was evident as they began to consider a species so familiar yet in so many ways unknown to them.

His class, Practicum in Field Ecology, was filled with eleven juniors and seniors, most planning on pursuing environmental careers. The class would enable them to gain firsthand experience in trapping and tracking wildlife. "I wanted the students to walk out the door and practice a skill in wildlife biology that is equivalent to what a state wildlife biologist, a federal wildlife biologist, or graduate student would do," said Perlut. Squirrels, specifically nonhibernating, native, diurnal gray squirrels, were the most practical for the undergraduate class. And so Project Squirrel began.

He's succeeded in reaching his goal. The students walk just steps away from their classroom to their outdoor campus research site. It had been an unusual year. There was only one gray squirrel still out in the wild with a collar until the week of my visit, when telemetry guided one of the students to an empty collar. The class theorized that the collared squirrel, named Tom Girl, fell prey to a red-tailed hawk that they had previously observed with a captured squirrel. Although Tom Girl was gone, the squirrel's data remained in the class research files. Delving into the data showed that she was a juvenile when caught and tagged on April 2, 2024. She was first named Tom. As the squirrel matured, the class discovered it was a female and added *Girl* to the end of her name. The computer screen was filled with location points the students had gathered on Tom Girl using telemetry since April. Her tag would be refurbished for use on another squirrel.

The disappointment of losing one of their subjects turned to excitement when the students found that two of their nine traps were filled. One trap held a chipmunk, which they released; the other held a gray squirrel, which would enter their research project. The male squirrel was weighed, measured, tagged, collared, and named Jerry before being released back into the oak forest. Jerry's story was beginning as Tom Girl's ended.

"I think a big thing about this class is that this project has been going on for a long time," said student Autumn Linden. "There's tens of thousands of points. I don't know if anyone has really utilized this data yet. And so there is this untapped database of just, like some of these squirrels were recorded for up to two years, and so a lot of us are not in the place to do anything with that research, but hopefully . . . one day some scientist will hear about it . . . this kind of data. It is such good data to have."

"One of the fun things about this class is that every class, we have someone pick out a research article . . . about squirrel studies that have been done," said student Rick Hamilton. "One of the big things that has been hammered home to us in this class is that not a lot of research has been done about squirrels and specifically gray squirrels, so it is one of those things that we are getting to study and investigate the habits of a species that nobody really gives any credit to. Like it's just like they're there. They get eaten. That's about it. But we're reading articles about their social communication structures, how they use both visual and auditory cues to talk to each other, how they react to other animals, how they react to predators. It is really fun seeing a more in-depth look at a species that so often gets ignored."

Hamilton added that for him, a main reason for taking the course was that he had interned at the Center for Wildlife at Cape Neddick, Maine, and was introduced to animal ambassador Skeeter the Gray Squirrel, who was "an absolute sweetheart." It got him thinking more about squirrels. "You see them everywhere and how they just kind of blend into the background, like songbirds or almost insects to a lot of people," who see them all the time. "They kind of just blend in. And it is just really interesting to think about the actual lives they end up living and how personable they can be sometimes."

In the years since Perlut began Project Squirrel on the campus of the University of New England, many of the unanswered questions about gray squirrels have been explored, but not for reasons you might expect or where you'd expect. While the students studied and collected data,

North America's ubiquitous resident rodent was becoming the subject of great concern and interest far across the ocean in Europe, beginning in England.

Throughout this complex history of America, native and nonnative squirrels weren't just engaging with park visitors and college students in America; they were being introduced into urban parks around the world, in countries including Australia, South Africa, Italy, and the United Kingdom, spawning additional stories that would shape our ecological and cultural history.

*Chapter Seven*

# As an Interloper

Sciurus hudsonius Pall.

> *"When Timmy and Goody Tiptoes came to the nut thicket, they found other squirrels were there already."*
> —Beatrix Potter, "The Tale of Timmy Tiptoes"

It is more likely that a student at University College London will see a gray squirrel than a native Eurasian red squirrel on campus. That squirrel familiarity on college campuses such as the University of New England and in North American city parks such as New York's Central Park might be welcome to a visiting American, but they also feel out of place, exotic, in a foreign city such as London. Seeing a gray squirrel in a London park is like spotting an invasive mongoose in Hawaii. The mongoose might be a familiar sight to a visiting Indian, who resides where the mongoose is a common native species, but not to native Hawaiians. Nonnative mongooses were brought to the Hawaiian Islands from India in 1883 to control rat infestations. Instead of limiting their diet to rats, they found native mammals, reptiles, birds, and other species a bit tastier. It became an example of a welcome import gone awry. Nonnative squirrel imports, like the gray squirrel to Great Britain, have also taken an unsavory turn as they displace native squirrels.

Worldwide, squirrel introductions immediately positively affected city residents, but the effect on their descendants is far from positive. These nonnative interlopers created chaos. It's one thing to introduce a native squirrel to a nearby city, as in the case of the squirrels brought from American woodlands to local cities; however, importing nonnative North American squirrels to European cities presents unforeseen challenges for native wildlife, including native squirrel species.

Great Britain's gray squirrel invasion is a prime example. In the scope of squirrel history, the intrusive population of gray squirrels now occupying all of England, most of Scotland, and all of Ireland east of the River Shannon is a symbol of a species that is "flattening diversity." Peter Coates focused his book *Squirrel Nation: Reds, Greys, and the Meaning of Home* on the struggle between the United Kingdom's native reds and the invasive American grays, which drew into the saga a host of lawmakers, British citizens, and celebrities including Phil Collins, Elton John, Richard Branson, and Judi Dench.

More than a century before the Lacey Act of 1900 in the United States prohibited the importation of all wild mammals and birds, squirrels began their exportation abroad to unwitting, albeit willing, Europeans. We generally point to founding father Benjamin Franklin as the first early American squirrel exporter. In early 1771 or 1772, Franklin wrote to his wife, Deborah, from England requesting that she send him some squirrels for his friends. "The Squirrels came safe and well. . . . A 1000 Thanks are sent you for them, and I thank you for the Readiness with which you executed the Commission," wrote Ben to his wife in January 1772. The truth was that a singular animal arrived, Mungo, called by the English by the nickname Skuggs. Unfortunately, Mungo escaped his cage and met up with a man and his dog, Ranger, causing his death.

"Alas! Poor Mungo!" wrote Franklin in a letter to Miss Georgiana Shipley on September 26, 1772, upon the squirrel's untimely death. "I lament with you most sincerely the unfortunate end of poor Mungo. Few squirrels were better accomplished, for he had a good education, travelled far, and seen much of the world. As he had the honour of being, for his virtues, your favorite, he should not go, like common Skuggs, without an elegy or an epitaph."

Franklin wrote a free verse of twenty-two lines eulogizing Mungo that conjured the wilds of America for readers abroad. The dead squirrel became a metaphor for England and her colony, revealing Franklin's early reluctance to revolution. Fortunately for Georgiana, she received a second Skuggs from Franklin and his wife that grew "fat and lively" and enjoyed "as much liberty as even a North American can desire." But Franklin wasn't alone in sending squirrels abroad, and many of the others survived to bear young and populate the countryside. Alas, Mungo wasn't the sole squirrel to travel far, see much of the world, and cause disruption.

A less famous squirrel was exported from America, sent to Paris from New Orleans by a Capuchin priest. After three days abroad, the priest gifted it to the French queen. According to a note on an illustration of the squirrel, the little American immigrant aroused much curiosity and amusement at the French court, where "its favorite thing was to jump onto the ladies' necks and hide." The note explained, "He would fly from one end of the room to the other and could not fly further than that. His tail was the most peculiar."

However, those few squirrels brought to Europe didn't match the disruption that was caused by the exports that followed. Before 1930, there was a minimum of seven introductions to the British Isles from North America. The most influential was the release of ten gray squirrels at Woburn Abbey in Bedfordshire, England, in 1890. The 11th Duke of Bedford, Herbrand Russell, imported gray squirrels from New Jersey. The small North American squirrels arrived in a land as foreign to them as it would be for any human, with different sights and smells. The squirrels could probably smell the red and fallow deer that inhabited the grounds even before they saw the strange animals, which were unlike the white-tailed deer familiar to them in New Jersey. The duke didn't stop with importing wild squirrels to his estate. He imported Chinese water deer and Reeves's muntjac deer in 1894, making the new habitat even stranger for the squirrels. But gray squirrels have been found to be resilient, and these were no exception. They adapted to their new ecosystem with its different food sources and strange inhabitants. Those initial ten rodents led to many descendants that were later given as gifts across at least seven sites in the United Kingdom and Ireland, solidifying their place in the British Isles. But there were more.

Another one hundred gray squirrels, known outside of North America as grey, were released in 1902 in Richmond Park in Surrey, and another ninety-one were released over the next few years in Regent's Park in London. Without recognition of the danger they would pose to the ecosystem, more releases followed. The last documented squirrel introduction was in 1929, when two were set free in Staffordshire's Needwood Forest.

But it was too late. The import problem was obvious, eventually leading to the Grey Squirrels (Prohibition of Importation and Keeping) Order issued under the Destructive Imported Animals Act 1932. Sadly, the damage was already done. After ten thousand years of having the British Isles to themselves, the native Eurasian red squirrel population was coming to terms with their invasive cousins. Gray squirrels in Britain are now well established, having reached an estimated 2.7

million in 2018 as the native Eurasian reds continued their decline. The small reds had already faced habitat loss and hunting, but the gray immigrants posed additional threats.

Eurasian red squirrels, traditionally throughout Europe and parts of Asia, have fur colors ranging from pure red to pure black. The Eurasian reds have various subspecies, with Great Britain being home to a population of pure reds, while southern Italy is home to populations of pure black squirrels. Eurasian red squirrels and American red squirrels are different species and differ both physically and in behavior. American red squirrels, most abundant across northern North American boreal coniferous forests that contain abundant conifer seeds, interlocking tree canopies, and cool, moist environmental conditions, do not have tufts on the tips of their ears like Eurasian red squirrels. In addition, Eurasian reds are not territorial like their highly territorial American cousins. Neither of these red squirrel species hibernates, although some squirrel species, like the thirteen-lined ground squirrel (*Ictidomys tridecemlineatus*), do. Most squirrels remain active during winter; some go into a hibernation state called torpor, and others hibernate but remain in a drowsy, foggy state. The significant difference between the two red squirrel species is how they store their food supplies. Both stash and store food for cold winters, but Eurasian reds scatter their food caches in multiple forest locations, while American reds create one massive midden they fiercely defend. One American red midden consisting of thousands of acorns was revealed within a metal telephone pole in Milwaukee, Wisconsin. The nuts poured out of the pole like tokens from a Las Vegas slot machine. Those large middens don't exist in the United Kingdom.

Some midden-building American red squirrel mothers split up their stash into middens they gift to their offspring during lean years. Talk about nepo babies! Discussed in a 2015 study, this "midden-gifting" behavior indicates that some successful squirrel moms have the foresight to know that the following year will bring more nuts for their young to gather. This increases the chances for their American red young to survive during harsh winters. It also translates to the genes of these thoughtful moms being passed down through family generations.

Grays in the United Kingdom have the upper hand over reds for many reasons. In addition to being slightly larger and heavier than the Eurasian red squirrels, the grays have an advantage by emerging from hibernation a few weeks earlier than their European cousins. The earlier emergence provides the invasive grays with more time to prepare for winter and breed in the summer. While both are arboreal and diurnal, grays eat seeds and nuts before they are fully ripened, which also allows grays to get to the food first and deplete the supply before the reds can. Grays locate their caches using spatial memory and olfactory cues. One hectare (about two and one-half acres) of British woodland can support one Eurasian red, whereas the same woodland can support any number from three to thirty gray squirrels.

This struggle between squirrel species led author and poet Jane Yolen, known as America's Hans Christian Andersen, to write her middle-grade novel *Trash Mountain* in 2015. Yolen lives part-time in Scotland, where she saw how gray squirrels impacted the native red squirrel population. "I knew quite a bit about the war between the red squirrels and gray squirrels," she wrote, "mostly from British magazines and newspapers." Her novel tells the tale of a curious young Eurasian red squirrel named Nutley who feels the squeeze from the "larger, faster, and more aggressive" grays. Nutley reaches out the "Paw of Friendship" to the gray squirrels, who respond by tossing the little red off a cliff. Taking an even darker turn, as nature often does, the grays subsequently kill Nutley's parents. Yolen wrote to her young readers, "This you should know: In the real world, Red Squirrels have been pushed out, marginalized, by the bigger, stronger, more aggressive, and (some say) smarter Grays."

Are the grays smarter? That's an additional need for study. Roughly seventy different definitions of intelligence are found throughout the scientific literature, according to the *Atlas Obscura* article "Is a Squirrel Smarter than a Fifth-Grader?" Some studies suggest that gray squirrels are smarter than believed. Yolen has done some research and questions the grays' caching behavior in her book.

Whether they are considered smart or not, the invasive grays are dominant in many areas. It has been found that they'll attempt multiple tactics to open a locked box, and they do actually remember the location of their caches, but more on that later.

Two studies over a two-year period took place in Kirby Park in Wilkes-Barre, Pennsylvania; at Central Connecticut State University and Stanley Quarter Park in New Britain, Connecticut; and on the campus of Smith College in Northampton, Massachusetts. The research team observed 255 caching events in Pennsylvania and 49 at the Connecticut and Massachusetts sites. All of the observations were made in areas of maintained lawn, near or under deciduous trees, or in woodlands. To the research team's knowledge, it was the first study to show evidence of behavioral deception by a rodent: "When scatter hoarding a food item, gray squirrels typically first excavate a shallow pit by digging with the front paws, then, with the food item in their mouths, push the food item into the base of the pit, often with several obvious thrusts of the entire body, and finally cover over the site by dragging the paws along with soil and debris towards the body (<10 times). Covering often also includes 'patting' movements in which the front paws alternate to tamp down the soil, and squirrels may end a caching sequence by combing vegetation and debris towards themselves to cover the cache site further. The entire process usually takes less than 60 s[econds]. However, to cache a single item, a squirrel may excavate several sites before deciding on a suitable cache site."

Another study explored the squirrel's use of "deceptive, or paranoid, behavior," pretending to bury a nut when being watched. The squirrel will dig a hole, pat it down with its front teeth, and scrape dirt or grass material over the top. But at the same time, it has concealed the nut near its armpit, enabling it to bury it elsewhere later. Like other behaviors that are often considered human, these perceived social identities, traits, and behaviors often reflect our own humanness.

In research conducted at the University of California, Berkeley, students were placed in a competitive game to behave like wild squirrels. Plastic eggs were hidden by the students, and fifteen minutes later, the students were asked to find them, replicating nut caching. But, unlike squirrels, most students couldn't remember their hiding places. Imagine burying ten thousand plastic eggs, the number of nuts squirrels bury each year. And then imagine attempting to remember these sites while avoiding predators and adverse weather.

The findings of squirrel deception in cache protection strategies have inspired additional studies. Researchers identified four traits that golden-mantled ground squirrels exhibit in their personalities that are familiar to humans: boldness, aggressiveness, sociability, and activity level. Squirrel researchers from Sweden found that participants described squirrels as attractive, charming, fun, lively, and alert.

Yet another study suggests that the invasive grays are preadapted to cope with the challenges of population density and other characteristics of urbanization. Is it just that gray squirrels can deal with the stressors of city life better than reds? Do these behaviors give gray squirrels an edge over beloved native reds, which even found their way into Shakespeare's writing?

An early squirrel mention depicting the anthropomorphically industrious squirrel is found in Shakespeare's *Romeo and Juliet*, in Mercutio's description of the chariot of Mab, Queen of the Faeries:

> *Her chariot is an empty hazel-nut*
> *Made by the joiner squirrel or old grub,*
> *Time out o' mind the fairies' coachmakers.*

But perhaps better known than Shakespeare's brief squirrel mention are the squirrels from the tales of Beatrix Potter, a mainstay of British children's literature. Her 1903 "tale about a tail" of a little red squirrel named Nutkin influenced generations of British schoolchildren. Even today, fans of the author's stories can enjoy a stage filled with dancing red squirrels performed by ballet dancers. The ballet was adapted for stage in 1992 from Frederick Ashton's full-fledged 1971 film production of the *Tales of Beatrix Potter*. Those dancing squirrels, including Matthew Hart's performance as Squirrel Nutkin, now fill social media reels.

Potter's *Tale of Squirrel Nutkin*, published over 120 years ago, was inspired by the native Eurasian squirrels she observed in Lingholm, in England's Lake District, just about twenty-five years after gray squirrels were introduced into Britain. Potter was a keen observer of animal behavior. She combined her scientific knowledge and observations with wit and whimsy to create the beloved characters in all her books, including Squirrel Nutkin. "And to this day, if you meet Nutkin up a tree and ask him a riddle, he will throw sticks at you, and stamp his feet and scold, and shout—'Cuck-cuck-cuck-cur-r-r-cuck-k-k.'"

As Potter's popularity grew, so did her stories. Almost a decade after her tales about British wildlife, she had an international audience, which led her to write a tale focused on gray squirrels for her American audience. "The Tale of Timmy Tiptoes" featured wildlife her American readers would recognize, including gray squirrels, chipmunks, and a black bear. By the time of its release in 1911, gray squirrels were becoming well established in Britain.

A little over a decade later, Felix Salten, the author of *Bambi: A Life in the Woods*, focused his book *Perri: The Youth of a Squirrel* on the fictional life of another Eurasian red squirrel.

The real-life Eurasian red squirrel (*Sciurus vulgaris*) counterparts to those fictional protagonists continued for years to struggle to survive despite efforts to control the interloping gray squirrels. One effort to reestablish reds involved releasing eight reds in 2016 in an area near Wales's Ogwen Valley. Although there have been some subsequent sightings, these reds never seem to be able to establish a population. Their struggles appear to point to losses from the squirrelpox virus (SQPV) spread by the grays.

Squirrelpox cases brought on by the Leporipoxvirus manifest in red squirrels as oozing sores or scaly growths or tumors that appear around the squirrels' eyes, mouths, and paws. The deadly, contagious virus isn't limited to the United Kingdom. It is currently spreading in the state of Maine and has also been found in the Lower Peninsula of Michigan and in Northern Indiana, where it has been found in red, gray, and fox squirrels. Although pox is seen in both the United Kingdom and the United States, it exists solely in the presence of gray squirrels. It is spread from squirrel to squirrel but can also be spread by mosquitoes.

Although the IUCN Red List of Threatened Species records their current population trend as Unknown, in 2018 Eurasian red squirrel numbers were estimated to be a mere 287,000. This low number is attributed to several factors, including competition with gray squirrels for resources, diseases such as squirrelpox spreading from grays to reds, and loss of habitat.

This dangerous relationship threatening the population of Eurasian reds has, fortunately, spurred a great amount of scientific examination that otherwise might not have been explored.

While students at the University of New England collect data on their local squirrels, students at University College London are involved in researching their native red squirrels, using camera traps and educating younger schoolchildren about their conservation. But, as we've seen, the research has moved beyond university campuses.

Lisa Signorile, from the Zoological Society of London's Institute of Zoology, dipped into the squirrel population to uncover the genetic roots of British gray squirrels. "Grey squirrels are not as crazy invaders as we think—their spread is far more our own fault," she said. Her 2016 analysis of 381 squirrels from the collection of the Natural History Museum in London and modern populations showed a population mosaic. Whereas the gray squirrel population was expanding in the years leading up to 2018, when their population was estimated at 2.7 million, Britain's native red squirrel numbers had been in decline since the 1940s.

Groundbreaking research was published in the *Journal of Medical Microbiology* in early 2024 that explored a significant difference in the diversity of gut bacteria between the two competitive species that points to another reason the gray squirrels can outcompete the smaller reds. It raises the saying "listening to your gut and changing behavior" to a new level. Are squirrels listening to their gut? Researchers collaborating from the Woodland Trust, University of Surrey, University of Bangor, and the Animal and Plant Health Agency sampled bacterial DNA from the gut contents of both species. They performed DNA sequencing to identify the bacterial groups that were present in the two squirrel species. They found that the gray squirrels had a more diverse range of microbiota in their gut. How did that finding correlate to success for the grays?

"Red squirrels are now an endangered species in the UK. Not only are grey squirrels larger and more robust than red squirrels, we have now identified a significant difference in their gut bacterial microbiota, potentially giving them another advantage over reds," Roberto La Ragione, professor of veterinary microbiology and pathology at the University of Surrey, explained in a press release.

What does all this mean? The study noted that the microbiota "is formed by communities of micro-organisms including viruses, protozoa, fungi and bacteria, and there is great potential for microbiome research to improve conservation outcomes." The increased gut microbiota in gray squirrels could indicate that they have stronger general health and immunity than their red cousins. According to the study, it could "influence the susceptibility of red squirrels to squirrelpox virus and adenovirus as well as potentially negatively affect the native sciurids' capacity to compete with" nonnative gray squirrels. It could also reflect the grays' broader diet. These differences provide researchers with valuable data on how gray squirrels can access a broader range of resources.

In addition, researchers found the presence of oxalate-degrading bacteria in the guts of gray squirrels, demonstrating that the squirrels are able to digest calcium from tree bark. This could be a possible explanation for their bark-stripping behavior, which damages trees and causes economic loss.

The dilemma is being studied throughout Britain, resulting in many organizations working together in their research. A joint report by the Royal Forestry Society, the Forestry Commission, Natural Resources Wales, the National Forest Company, and the Woodland Trust in 2021 estimated that gray squirrel damage to trees could cost at least £1.1 billion over the next forty years if the population of gray squirrels continues to flourish in the United Kingdom. The importance of the commercial impact on the native forests in the United Kingdom can be seen by the length and position of the reasons to control gray squirrels in a 2021 fact sheet published by UK Squirrel Accord and the Animal and Plant Health Agency, where merely two sentences on that impact on the native red squirrel population followed the large first section of the effects on the UK treescape, and just one sentence focusing on the grays' impact on songbirds followed as the third and last reason.

Just a year before this fact sheet was produced, more than two hundred personalities took part in Scribble a Squirrel, an effort organized by the Northumberland Wildlife Trust, Red Alert North East, and the Calvert Trust to support red squirrel conservation. Each celebrity, including Paul McCartney, Heather Mills, and Michael Caine, drew a squirrel, and the drawing was auctioned off during Red Squirrel Week. McCartney's "Pink Squirrel," inspired by Squirrel Nutkin, received

£4,120, the highest price. He followed with other squirrel scribbles, finishing up with a red squirrel that became the protagonist, Wirral, in his first children's book, *High in the Clouds* (2005). Wirral escapes to a utopian sanctuary where all creatures live in harmony. The squirrel eventually falls for a squirrel named Wilhelmina, reflecting a hopeful red–gray coexistence. But in reality, the coexistence of these two squirrel species is exigent.

In *Squirrel Nation*, Coates pointed out that native red squirrels were never known as reds until the grays ventured into the countryside; they were known solely as common squirrels. They were beloved but also characterized as pesky "tree rats" that ravaged the nests of songbirds and destroyed tree saplings, attitudes toward native squirrels that reflected a natural ebb and flow of feelings. This phrase reflects the natural fluctuation in human attitudes, much as with the grays in their homeland. Attitudes shifted when the squirrel population was threatened and the common squirrels became less and less common.

While this newest research is timely and crucial in delving deeper into the association between the native Eurasian red squirrel population and the invasive grays, it also highlights the need for more research. More studies are required to uncover the gut microbiota of other native red populations across Europe and Asia. This information is crucial in protecting other squirrel populations. As with any research, more questions develop. Where could this lead? Perhaps researchers can use these findings to develop methods to promote healthy gut bacteria in native red squirrel species, or explore the gut bacteria of endangered squirrel species outside of Europe. The need for further research is clear, and it can be an exciting time to be involved in squirrel research. Perhaps scientists will discover ways of preventing tree damage that don't include culling squirrels.

Ultimately, these studies could also help preserve the health of native squirrel populations in the United Kingdom and beyond and promote a greater understanding of all squirrel species. In addition, researchers stressed that antimicrobial resistance is a real threat to human and animal health globally. The previously mentioned analysis of animal microbiota diversity "also has implications for human health as wild birds and mammals potentially harbour and transmit zoonotic bacterial pathogens and associated antimicrobial resistance (AMR) due to their free-ranging behaviours, feeding habits and ability to come into close contact with humans and livestock."

While the competition between these two squirrel species is driving much of the research in Europe, other researchers and conservationists are exploring all options to conserve the native red population. As with other threatened species, science and conservation go hand in hand.

Even artificial intelligence (AI) technology is being utilized. Project Red Haven is deploying Genysys Engine in the United Kingdom to automatically detect and identify red squirrels, pine martens, gray squirrels, and squirrelpox in living environments to enable population monitoring so that conservation strategies can be tailored to assist conservationists.

Perhaps one of the most drastic methods of conserving reds was the decision to put those invasive gray squirrels on British menus. "When the grays show up, it puts the reds out of business," said Rufus Carter, managing director of the Patchwork Traditional Food Company in Wales. Patchwork is just one business that added squirrels to its offerings, offering a squirrel-and-hazelnut pâté on its menu.

Britain launched the Save Our Squirrels campaign in 2006 to rescue its threatened red squirrels by piquing the appetites of Brits for their interloping North American gray cousins. This campaign wasn't the first time the public was encouraged to dine on one species to save another. Michelin-starred chef José Andrés has served up invasive lionfish to contribute to coral reef conservation in the Caribbean, and feral hogs have appeared on menus in America. This concept of "invasivorism" was developed over two decades ago at the University of Vermont in the belief that humans are tremendous predators and can function as a form of biological control. A website, EattheInvaders.org, was developed to encourage the eating of invasive species by sharing recipes. The site's tagline is "Fighting Invasive Species, One Bite at a Time."

Across Britain, the motto "Save a red, eat a gray!" created a market for culled gray squirrel meat. The new menu item had become so popular in the United Kingdom that it was featured in a 2009 article, "Saving a Squirrel by Eating One" by Marlena Spieler for *The New*

*York Times*. "These days, however, in farmers' markets, butcher shops, village pubs and elegant restaurants, squirrel is selling as fast as gamekeepers and hunters can bring it in," wrote Spieler.

Restaurants in Britain added the meat to the menu. Squirrel tastings sprang up. Nichola Fletcher, food writer and co-owner of a venison farm, provided a spark for squirrel flavor by hosting a squirrel tasting for Britain's Guild of Food Writers. At first, Fletcher claimed, "their lovely flavor tasted of the nuts they nibbled," but she later admitted she found the meat had "a greasy texture and unpleasant taste."

Britain's campaign to protect native red squirrels by eating nonnative gray squirrels also suffered from protests by animal rights activists. Protesters menaced restaurants that included squirrels on their menus. Animal rights activists went so far as to threaten to firebomb the historic Hadley Bowling Green Inn in Droitwich, Worcestershire, and smash the staff's cars over the restaurant's squirrel pâté appetizer after the dish was featured on a local television episode in 2009. The historic pub, restaurant, and inn have seen a lot over the past four hundred years, but pressures to pull squirrel meat from its menu escalated that year and added a troublesome new chapter to its long history. Those threats from animal rights activists succeeded in removing the starter from the menu.

But the movement continued. "While some have difficulty with the cuteness versus deliciousness ratio—that adorable little face, those itty-bitty claws—many feel that eating squirrel is a way to do something good for the environment while enjoying a unique gastronomical experience," wrote Spieler.

Danny Kingston, known online as Food Urchin, wrote about "helping people overcome their prejudices" in a recipe on Great British Chefs' blog for Potted Squirrel with Sourdough. The comments he received included "How could you?"; "What? Like the ones we see in the park?"; and "So, you are eating rodents now are you Dan?" He isn't sure whether he convinced his mother, who left many of the comments, but he still urges people to try squirrel. It's a "lovely, sweet, alternative source of protein to eat. I know Pascal Aussignac is a fan of this sustainable meat and they are becoming a lot easier to come by."

King Charles, former patron of the Red Squirrel Survival Trust as

Prince of Wales, long supported the effort to conserve Britain's red squirrel population. He stated, in honor of Red Squirrel Appreciation Day, January 21, 2021, "As you all know so well, these charming and intelligent creatures never fail to delight. . . . They are such inquisitive and delightful characters; they have even been known to hunt down a few of their favourite nuts left out in an unguarded jacket pocket! . . . I am so very grateful to all of you, as volunteers, for the crucial role you play in this ongoing battle to protect and restore a precious part of our natural heritage." After Charles became king of England, new coins featuring several native species across the British Isles were minted to reflect his interest in conservation. Among the wildlife represented was the threatened native red squirrel on the twopence.

Robert Gooch, owner of the Wild Meat Company, has been selling gray squirrel meat for over fifteen years and sees it as the perfect sustainable product. Looking beyond the culling aspect to protect reds, he notes that gray squirrels, like all wild game, are absent from human intervention and energy consumption, which is found in farmed meat production. Gooch asserts that unlike the other wild game he sells, the squirrels he carries are "humanely" trapped instead of shot by hunters. Many see it as a viable option. The company's squirrel meat description compares the taste of squirrels to that of rabbits, rather than the standard comparison to chicken. The company says the meat is "finer textured and has a more subtle flavour" than rabbit meat.

Head chef and owner Doug McMaster of Silo, a zero-waste, Michelin Green Star–awarded restaurant in east London, hosts an invasive species supper club where he dishes up invasive species including crayfish, Pacific oysters, fallow deer, and gray squirrels in elaborate menu items such as fallow deer parfait, crayfish tartlet, and gray squirrel kofta. "We choose food sources that respect the natural order, allowing ingredients to be themselves without unnecessary processing," he says. McMaster has won awards for his innovative restaurant, including Britain's Most Ethical Restaurant. He strives to "creatively popularize species that are detrimental to the environment." The invasive species he serves at his restaurant, he claims, are all "'forces of destruction' that squeeze out local populations, but all are edible and 'delicious.'"

However, harvesting and serving grays on British menus is just one potential answer to native red squirrel recovery. British researchers have also moved red squirrel conservation efforts in another direction, boosting the population of gray squirrel predators in England's northern Lake District, opting instead for a nonhuman predator to devour the grays. Staff at Skelwith Fold Caravan Park planted dozens of Scotch pines to form an ideal habitat for European pine martens (*Martes martes*), a native cat-size member of the weasel family. The pine marten population in Britain has also suffered from habitat loss and hunting. Conservationists see the recovery of the pine martin as a win-win for the UK environment.

The first conclusive sighting of a native pine marten in over one hundred years occurred in the summer of 2015. The squirrel predators were virtually extinct before this tree-planting recovery effort. Experts hope that returning the squirrels' predator to the ecosystem will help naturally cull the invasive gray squirrels that have never encountered pine marten predators.

Eurasian red squirrels coevolved with pine martens, according to Emma Sheehy, one of the authors of a scientific article published by The Royal Society in March 2018. "They've managed to co-exist over such a long time and their population doesn't seem to be affected by losing the odd individual to a predator."

Native European pine martens still have a strong population in Scotland, where Sheehy has studied their relationship with invasive grays. Her inquiry suggests that pine martens reverse the "typical relationship" between reds and grays, in which reds always lose. "Where pine marten activity is high, grey squirrel populations are actually heavily suppressed. And that gives the competitive advantage to red squirrels." She found that squirrel numbers increased in areas where they hadn't been in some time. "Pine marten predation thus reverses the well-documented outcome of resource and apparent competition between red and grey squirrels."

It is hoped that pine martens will reestablish themselves in northern England with the help of habitat recovery from tree planting in time to restore the naturally healthy predator–prey relationship that may ultimately provide an edge for the quicker native reds against the larger, slower grays.

Pine marten recovery in Northern Ireland has already been successful. Currently, fourteen groups there are working to preserve their native red population. Each team of volunteers goes into the woodlands and sets up squirrel feeders and trail cameras to monitor whether reds, grays, or pine martens are present. The conservationists also plant trees to build habitats and plan educational programs to bring awareness and encourage action to promote the environment and local biodiversity. But they don't stop with those measures. They work to remove and cull the gray squirrel population at the same time. One of the groups has removed about six hundred gray squirrels since 2011. These combined efforts have seen success. The volunteer efforts in restoring the pine marten in County Fermanagh have been so successful that the area is now regarded as free of gray squirrels. With that comes more welcome red squirrel sightings.

Culling the grays and introducing a predator are two of the substantive ways that are in play in Great Britain, but Brits are also exploring a red squirrel breeding effort underway at places such as the Yorkshire Arboretum, where staff recently rejoiced at the births of four red squirrel kits to two different native squirrel moms, Holly and Hazel. The program, though, needs help to work. Its success has been lower than desired. John Grimshaw, director of the arboretum, said, "Somebody told me at the beginning it was easier to get hold of a lion cub than a red squirrel." The male squirrel, appropriately named Erik the Red, who fathered the celebrated births in 2023 was the sole native red squirrel they were able to find for the program. And, as with any captive breeding recovery program, there is always the concern that the matches will not breed. For the program to have any future, the kits born at the arboretum will have to be sent to other breeding facilities to continue the recovery efforts by the British and Irish Association of Zoos and Aquariums (BIAZA) Red Squirrel Studbook network. This will ensure a genetically diverse captive population. While the red squirrel breeding program is another tool in the toolbox, researchers are also considering gray squirrel fertility control.

The Red Squirrel Survival Trust discusses another form of fertility control being explored by the United Kingdom's Animal and Plant Health Agency that involves developing an oral contraceptive for gray squirrels. Of course, this laced bait would have to be administered in a species-specific hopper that would be inaccessible to red squirrels and other wildlife so that the contraceptive effect is strictly controlled for grays. An injectable single-dose contraceptive, GonaCon, developed by the National Wildlife Research Center, has been used to manage populations of white-tailed deer, horses, and donkeys. The immunocontraceptive vaccine is registered in the United States and used by the US Department of Agriculture (USDA), Animal and Plant Health Inspection Service (APHIS), Wildlife Services (WS), for use in female wild or feral horses and burros, white-tailed deer, and prairie dogs (California ground squirrels). It is approved for fertility control of black-tailed, white-tailed, and Gunnison prairie dogs. GonaCon–Prairie Dogs is classified as a general-use pesticide for use by USDA, APHIS, WS, or state wildlife management agency personnel and must be kept away from humans, domestic animals, and pets. It is prohibited from direct use in water.

This offers an alternative to shooting and trapping, which can be expensive and cruel. In 2015, for example, Red Squirrels Northern England worked to cull 21,000 squirrels in northern England, which was estimated to cost £60 per squirrel. In 2024, the National Animal Rights Association (NARA) in Ireland contacted the National Botanic Gardens in Dublin, which had been using "horrifically cruel" traps to cull squirrels from the popular tourist site. The traps, constructed to resemble squirrel nesting boxes, contained a piston powered by compressed gas deployed when a squirrel peeked inside, striking the squirrel's head to cause a "near-instantaneous" death. However, officials admitted that animals hit in the trap sometimes lived for painful minutes after falling to the ground. Amid outcry from visitors and animal rights activists, the traps were removed. Are there other options to cull the gray squirrel population and preserve the native reds?

Using contraceptives, either alone or in addition to culling, might be the answer for preserving the Eurasian red squirrel population. The plan is claimed to also help officials reach the target goal of having 16.5 percent of England under tree cover by 2050.

As I write, the United Kingdom has been struggling with this issue for over ten years without success and could be facing the extinction of native reds in another ten years. That said, they have not exhausted all options to recover native reds. Other tools are being explored, including the use of clustered regularly interspaced short palindromic repeats technology, better known as CRISPR, to modify the DNA of gray squirrels so that they pass on infertility genes. This technology was developed in 2012 and makes it much easier than previous methods to edit the genomes of plants and animals. Using this method would ensure that just one sex could be bred, causing population decline due to female infertility. A study published in *Scientific Reports* in 2021 claims that releasing one hundred mutant squirrels into the wild in Great Britain could wipe out the population within fifteen years. The study maintains that CRISPR can "offer a humane, efficient, species-specific and cost-effective method for controlling invasive species, including grey squirrels in the UK." But that research was based on a computer model. It might be more challenging to see this through in the wild.

Award-winning filmmaker Terry Abraham's 2023 documentary, *Cumbrian Red: Saving Our Red Squirrels*, debuted in February and March 2024, coinciding with the launch of a new conservation project in Furness in the Lake District. The documentary was created after Abraham spent two years volunteering for the Penrith & District Red Squirrel Group. It highlights the plight of the native red squirrels within the Lake District National Park and Cumbria. Abraham was nervous about the film's London premiere because the city's gray squirrels are often the single connection residents have with nature. He acknowledged that the situation is the fault of the humans, not the grays, for introducing the nonnative species in the first place. "We don't want to demonise them, however now it really does rely on human intervention to save the reds," he said in a February 2024 article.

Sadly, most of the population he documented perished. He posted on X (formerly Twitter), "I'm honestly gutted. I'd see reds often at Rutter Falls. . . . I got to know the reds at beautiful Rutter Falls and the place features in my film 'Cumbrian Red.' I can't believe how quickly the reds have been wiped out. I'm gutted but bloody angry too that more isn't done to inform the wider public by big charities, local authorities and government about a legally protected, iconic and national favourite species."

In the meantime, other conservation efforts to preserve the Eurasian reds continue, and squirrels remain in the news when housing developments are curtailed to conserve native squirrel habitat.

Throughout the years of preservation, squirrels have continued to serve as "historical actors" on the international stage. The British dilemma made international headlines in 2023 amid the war in Ukraine as Russia's state-run television turned the conservation actions into a political message, claiming falsely that people in the United Kingdom, facing war-related food shortages caused by Britain's military aid to war-torn Ukraine, had been reduced to eating squirrels.

"Today it was revealed that some restaurants in once-Great Britain will be serving squirrels," said Olga Skabeyeva on Russia's *60 Minutes*, according to the translation by Ukrainian interior ministry adviser Anton Gerashchenko. "They [the UK] are not backing down from the decision to help [Ukranian President Volodymyr] Zelenskyy, to supply weapons. That is, they will eat squirrels, but still supply howitzers." Who knew red squirrel conservation could become a topic of Russian propaganda?

Anton Gerashchenko posted on X, "Russian propagandist Skabeyeva says that some restaurants in Great Britain started serving squirrels because of the food shortage in the country. She implies that all the money went towards weapons for Ukraine, and now Britons must eat squirrels."

But gray squirrels, going about their squirrel lives, weren't aware that they would be in the news again later that year, when they were oddly compared to a terrorist organization during a session of the British Parliament.

On November 28, 2023, a month after Hamas's deadly terrorist attack on Israel, Jim Shannon from the Democratic Unionist Party described gray squirrels as terrorists during a Parliament debate on the control of the invasive squirrel population. In praising the volunteers working to save the native red squirrel population, Shannon said, "Indeed, the issue is the very presence of grey squirrels—grey squirrels are the Hamas of the squirrel world."

As the struggle between the two species continues in Britain, research also continues that will benefit Eurasian red squirrel recovery and preservation everywhere. The Welsh government, for example, is calling for funding to develop a squirrelpox vaccine that would help the existing populations of reds succeed. The vaccine is a much-needed tool in the toolbox to ensure the survival of red squirrels. As an example of how dangerous squirrelpox can be, a single pox-carrying gray carrying the virus to the Welsh island of Anglesey could wipe out the island's entire red squirrel population.

Currently, each Eurasian red squirrel sighting is a treasure as the gray squirrels continue to spread. There was even a "cheeky" gray squirrel found just a few feet below Trfyan's Adam and Eve summit stones of one of the most famous mountains in Eryri (Snowdonia) in Wales in the summer of 2023. The little rodent was perfectly camouflaged against the mountain peak's tumble of gray rocks. But this wasn't the first time a gray squirrel was spotted on Eryri's summits; one was seen in 1975 and another in 2012, and one was claimed to have snuck into Wales's highest building and stolen a packet of nuts.

Conservationists are even employing artificial intelligence to conserve red squirrels. Scientists already use AI to spot puffins on Scotland's Isle of May. Squirrel Agent: Project Red Haven is a program that takes the AI method further. Developed by Genysys Engine in Ireland, the program can detect different species of squirrels using physical markers such as chins, ears, and tails. When the system spots an invasive gray squirrel, a push notification appears on the phones of conservationists, who can deliver contraceptives through a triggered trap or squirrel feeder.

All of these concerns and solutions continue to serve as inspiration and tools for protecting native species around the world as invasive species spread during a period of global change that is already challenging the survival of native wildlife species. The use of a contraceptive does come with the risk that humans or other wildlife might eventually consume a treated animal, though APHIS and the USDA claim this is unlikely and negligible: "Direct risk to nontarget fish and wildlife and any impacts to habitat or prey items (indirect risk) would not be expected to occur under the proposed use pattern for GonaCon."

Each plan to conserve Eurasian red squirrels in Britain has many considerations to explore. Regardless, a solution is vital. Ultimately, if efforts are unsuccessful, the United Kingdom could face the extinction of native red squirrels. Professor Karen Munroe of Baldwin Wallace University also fears the possibility that gray squirrels would eventually cross the Alps, spreading to Asia. Gray squirrel spread would endanger an increased number of native squirrel species populations across the globe.

Many conservation organizations strive to rein in the population of invasive nonnative squirrels and manage the existing European populations so that doesn't happen. But even now, grays are already on the European continent, without crossing the Alps or swimming across the English Channel. As with the grays' invasion of Britain, somebody brought them there.

While the grudge match between gray and red squirrels made international news and created a conservation conundrum in Britain, gray squirrels were becoming a common sight in parks in northern Italy.

*Chapter Eight*

# On the Continent

Plate 100

SQUIRREL.

> *"For us to go to Italy and to penetrate into Italy is like a most fascinating act of self-discovery."*
> —D. H. Lawrence

> *"How beautiful is sunset, when the glow / Of heaven descends upon a land like thee, / Thou paradise of exiles, Italy."*
> —Percy Bysshe Shelley

"Everywhere in Torino [Turin] it is possible to see a gray squirrel," said Italian scientist Sandro Bertolino during my visit to the northern Italian city. "Their ability to adapt is incredible." We sat in his university office, rain pouring outside the window and a wooden squirrel statue celebrating Bertolino's sixtieth birthday perched on a shelf. Indeed, on the previous sunny October day it had taken me only about ten minutes to see an intrusive nonnative gray squirrel on the tree-lined path in Torino's oldest park, Parco del Valentino, on the west bank of the Po River. The familiar squirrel looked at me from a tree trunk and then scrambled over in hopes of food—a quick indication of how many human residents share their snacks with the friendly rodents. It was all too familiar. I could have been home in New York's Hudson Valley. Just like the grays seen in London parks, this one spotted in Italy was out of place.

The reception to nonnative squirrel species in Italy is mixed. The friendly invasive gray squirrels are well-liked in the city parks. After all, most of the urban residents weren't accustomed to seeing red squirrels regularly before the introduction of the grays, so they aren't feeling pangs of loss. "Only people with large property miss them," explained Bertolino about the shy native red squirrel population.

A few days later, I spotted another gray squirrel among the hazelnut trees in the countryside of San Sebastiano da Po, about forty minutes outside Torino. The Po River, explained Bertolino, has become a corridor for their movement since the squirrels' introduction.

As in Britain, human intervention caused the problem with invasive squirrel species in Italy. Humans imported the eastern gray squirrel (*Sciurus carolinensis*), Finlayson's squirrel (*Callosciurus finlaysonii*), and Pallas's squirrel (*Callosciurus erythraeus*) to the European continent.

The squirrels seen today are descendants of earlier introductions.

The gray squirrels have established a new home in northern Italy, currently the sole country in continental Europe where they are found, after being introduced into the Italian province of Piedmont (Piemonte) in 1948 by careless Italian diplomat Giuseppe Casimiro Simonis Vallario, who enjoyed the "exotic" animals while visiting Washington, DC, for meetings following World War II. He returned to Italy with four little squirrels, which he released into a park near his Torino villa. The squirrels eventually escaped into the wild and colonized the nearby woods of Stupinigi.

Bertolino sat with me in his university office in Torino, not far from Bosco di Stupinigi, the site of the first gray squirrel introduction into Italy. He is one of the leading researchers on the invasive species in Torino and Varese.

Stupinigi is a hamlet, a *frazione*, of the commune of Nichelino, about ten kilometers, or six miles, from the center of Torino. It is best known for the Palazzina di Caccia, built as a hunting lodge in the early eighteenth century for the royal House of Savoy. The woodlands surrounding it, Parco di Stupinigi, are now protected and provide a vital environment for many wildlife species, including weasels, foxes, dormice, hares, and native red squirrels. However, the park's website now states that "some exotic species . . . by competing with indigenous species, are causing their disappearance. The most striking example is that of the American [eastern] gray squirrel (*Sciurus carolinensis*), which has led to the local disappearance of the European red squirrel (*Sciurus vulgaris*)."

In 1966, about a decade after the introduction of the initial gray squirrels, five more squirrels were imported from Norfolk, Virginia, and released into the Villa Gropallo Park at Genoa Nervi. Like Stupinigi, Parco di Villa Gropallo welcomes visitors to the former sixteenth-century villa gardens surrounded by sea and busy roads. One local tour website highlights the "best loved by children, semi-tame yet curious squirrels" as an attraction but doesn't mention that the squirrel species in this "Eden" are nonnative grays.

More invasive squirrel introductions followed, including one in 1994 when the municipality of Trecate, an industrial commune in the

Piedmont Province of Novara, funded the release of three gray squirrel pairs. The pressure to eradicate these pairs led to their capture two years later. An accidental release in Perugia, Italy, in the early 2000s instigated a new invasive squirrel population in central Italy. These introductions, and others, set up the native Eurasian red squirrel populations for increased challenges.

If only it were the sole invasive squirrel species, it might be a little easier to control, but North American gray squirrels aren't the only nonnative squirrels that have found Italy, specifically Piedmont, hospitable. Italy currently has five squirrel species: the native red, Calabrian black, and three new introductions and three nonnative introductions. The native Eurasian red resides in the Italian peninsula with three subspecies, missing only from Salento and the Italian islands, including Sicily. The Eurasian red is typical in both the Alps and the Apennines. While it faces competition with the invasive gray in the northern Italian provinces of Piedmont and Liguria, it also suffers from habitat fragmentation.

"Squirrels proved to be successful invaders and their importation should be restricted so as to prevent further introductions," wrote Sandro Bertolino in the conclusion of "Animal Trade and Non-Indigenous Species Introduction: The World-Wide Spread of Squirrels," published in the journal *Diversity and Distributions* in 2009.

Along with the gray squirrel, Piedmont has seen numbers of Finlayson's squirrels (*Callosciurus finlaysonii*), which are native to Cambodia, Laos, Thailand, Vietnam, and Myanmar. This squirrel, not uniquely competitive with reds in these countries, has caused dramatic damage to vegetation in Italy. Like many other squirrel species, the Finlayson's squirrel entered Italy for ornamental purposes. It arrived in southern Italy in Maratea, in the Basilicata region, in the mid-1980s. The nonnative Finlayson's squirrel then spread north and south along the Tyrrhenian coast. A squirrel cloaked in gray fur with a pale yellow belly, it has now made its way to the southernmost area of Campania at a rate of roughly 1.4 kilometers (0.9 mile) per year, rapidly expanding inland, where researchers have collected the first evidence of black morph squirrel individuals. These are potentially attributable to crossbreeding with the Calabrian black squirrel (*Sciurus meridionalis*, Lucifero 1907), a near-threatened and little-studied black tree squirrel with prominent ear tufts that is endemic to southern Italy. The black squirrel, initially considered a native red squirrel subspecies, was found to be a separate species in 2017. Recent studies of the squirrel's movement might indicate an ecological corridor that would provide a means for strengthening nonnative dispersal. Although it has been identified as a priority for eradication, no action has been implemented to prevent further spread of the black morph, not even in protected areas. Researchers call out the urgent need for studies to improve the small bank of knowledge of the Calabrian black squirrel and identify the primary variable that might compromise its survival so that researchers can consider management strategies. Among the needed strategies is the means for effective containment of the interloping Finlayson's squirrel.

Photographer and postgraduate student Francesca Sironi lives near Milan, Italy, and frequents her nearby parks with her camera. She's taken an array of red squirrel photos, which she posts on Instagram. "Monza was an oasis for red squirrels until three to four years ago," she told me. "Now there are only five red squirrels, and no one controls them. It's full of gray squirrels." She added, "A couple of months ago, I discovered two red squirrels in another park in my city." She hadn't spotted any grays there in the past three years, but now she's found two gray squirrels building a nest in the red squirrel territory despite the invasive parakeets that attack the squirrels every time they move. "The female red squirrel was pregnant last month. I really hope that she and the babies are fine and not too stressed."

Unfortunately, Sironi didn't know of any conservationist organizations working on protecting the native red squirrels in her area, despite the studies that the universities are conducting. During my visit to the nearby mountains of Valchiusella, near Torino, I walked through forests of hazel trees and found the ground littered with nuts. According to my guide, Adam Rose, native Eurasian reds still inhabit the mountain area. "We see them. I saw one today squashed on the road. They are feeding on those hazelnuts. They are stashing them and caching them."

But the native reds are not tremendous in number, are shy, and spend much time in the forest canopy; hence, I didn't see a single one. The finding of a dormouse tail on the trail indicated that rodent predators are also present in the ecosystem and possibly prey on the reds.

In addition to the problems with other invasive squirrels, the news about grays continues and overshadows the studies. Reports show that grays appear on hazelnut farms, which can have an adverse economic effect if they munch on the bark of the profitable Italian crop. However, Lisa Signorile, part of a research team examining damage caused by American gray squirrels to agricultural crops, poplar plantations, and seminatural woodland in Piedmont, found that "little or no information is available on the damage this species is causing on woodlands or agricultural areas."

In her continuing research, Signorile and her team applied DNA profiling to investigate the European invasion of the American eastern gray and found illegal squirrel trade in Italy. In a 2016 paper published in *Biological Conservation*, their results "revealed precise details of illegal squirrel trade in Italy."

Since then, the gray squirrels have damaged commercial poplar plantations and displaced the native Eurasian red squirrel (*Sciurus vulgaris*) population from an area of roughly 350 square kilometers (135 square miles) of the Piedmont Po Plain. Researchers found that the density of squirrel dreys increased with tree species diversity in the Stupinigi forest.

The grays have breached the Piedmont province and entered the Liguria province. However, as seen, the gray squirrel is not alone in threatening native Italian squirrels. The other species introduced, Finlayson's squirrel, the Siberian chipmunk, and Pallas's squirrel (*Callosciurus erythraeus*), are also present, and competition continues among them.

Pallas's squirrel, a smaller red-bellied or cream-bellied tree squirrel without ear tufts, is a native of Southeast Asia that entered the global invasive squirrel club through the pet and ornamental trade, like the Finlayson's. It escaped from botanical gardens, zoos, and homes to enter France in the 1970s, the Netherlands in 1998, Belgium in the 2000s, and Italy in 2007. It has also been introduced into Argentina, Hong Kong, and Japan. Like the invasive gray, it outcompetes the Eurasian red squirrel for food and nesting habitats. It also consumes bird eggs, insects, and snails and debarks trees. Like the invasive grays, Pallas's squirrel successfully becomes established in the wild after introduction. Of the twenty-nine known introduction events, twenty-one initiated an established population. But the grays are still getting the most attention.

Not long after John Platt's explosive 2012 article in *Scientific American* titled "Italy Faces Invasion of American Killer Squirrels," the terrorist analogy appeared in the British Parliament. Platt's 2013 article, "Squirrels: So Bushy-Tailed, So Ubiquitous—So Deadly," didn't mince words, nor the truth; the gray squirrel invasion has been fatal as the squirrels outcompete the native red squirrels and transmit the deadly squirrel parapoxvirus, commonly called squirrelpox. At the time of the article, Platt noted that "some scientists estimate that red squirrels could be extinct in the U.K. in as few as 20 years."

The National Wildlife Institute, in cooperation with the University of Turin, created an action plan to eradicate Italy's population of American gray squirrels, but, as with the culling in Britain, the first step of eliminating the population at Racconigi in May 1997 led to opposition from animal rights organizations and legal action that caused the project's suspension. A 2003 paper by Sandro Bertolino and Piero Genovesi noted the spread and attempted eradication of the gray squirrel (*Sciurus carolinensis*) in Italy and consequences for the red squirrel (*Sciurus vulgaris*) in Eurasia. It explained that although the National Wildlife Institute's acquittal occurred in 2000, the three-year suspension of conservation efforts caused a significant growth in the squirrel's population and the probable spread further into continental Eurasia.

But the story of gray squirrel and red squirrel competition in Italy isn't complete. The invasive Pallas's squirrel, native to China, India, Taiwan, and Southeast Asia, has been found in areas where the gray squirrel isn't present, creating a competition solely between the native reds and Pallas's, which causes some confusion. The two squirrels look similar. However, grays have a white underbelly and a reddish-brown face, whereas the smaller Pallas's have a reddish belly with an olive-brown face and back. Both the gray and the Pallas's lack the ear tufts of the red squirrels. In a 2017 study published in *Biological Invasions*, researchers explored the competition between Eurasian reds and Pallas's for space and food resources in mixed deciduous forests in Italy. When Pallas's squirrels were removed, the red squirrels increased their home ranges.

In another study, Italian scientist Maria Vittoria Mazzamuto and fellow researchers noted that while interactions between European red squirrels and gray squirrels are widely studied, little is known about possible interactions among other native and introduced squirrel species in Europe. Researchers found that competition for food resources led to native red squirrels having poorer body condition, emphasizing a risk to native species. "Red squirrels occurred at much lower densities and showed reduced adult survival in areas of co-occurrence than in red-only sites, but there were no differences in reproductive rate." Their conclusion underlines the "importance of a rapid eradication of the IAS [invasive alien species] and how management can have relatively quick impacts at least on the habitat use and resource exploitation of native red squirrels."

Mazzamuto and I spoke between her scheduled trip to the United States and my trip to Italy. "Why squirrels?" I asked her. Most researchers have their own squirrel story that sparked their adult research, but Mazzamuto is from Sicily, where there aren't any squirrels. Unlike the few islands in Great Britain with small populations of Eurasian red squirrels and Japan's islands with populations of flying squirrels, Sicily is squirrel free. "We have dormice in Sicily, but no squirrels. . . . My very first interest in squirrels was when I started my PhD and realized there was this conservation issue between invasive and native squirrels. That's when I got hooked on squirrels." Her path highlights how the squirrel competition continues to draw increased squirrel research.

However, Mazzamuto's PhD research focused not on invasive gray squirrels but on a relatively new invasive squirrel, the Pallas's squirrel, introduced in northern Italy near the green city of Varese. The city of Varese, filled with old villas and tree-filled parks, is a perfect setting for a squirrel invasion. Mazzamuto set out with her research to show people what was happening with this other invasive squirrel species and to explore whether there was an issue with Italy's native Eurasian red squirrel population. "It was interesting to approach this problem from bare ground."

She took a step-by-step approach to her research as she explored more about the Pallas's squirrel and its impact on Italy's native wildlife. "There is so much morphological variability," claimed Mazzamuto. "The variability in fur is between populations from different countries, but the animals in Italy are all the same, brown-olive back and cream belly." This makes the identification work more vital. The French Pallas's population appeared with red bellies, which was typical, but the ones in Italy exhibited cream abdomens. Mazzamuto needed to compare the Pallas's populations in France, Belgium, and Italy, including both the wild populations and those in museum collections to identify species accurately. She delved into and compared their morphological measurements, conducted DNA results, and compared their body colors to determine the identification and origins of the new invasive squirrels in Italy.

After identification of the species that had found its way to Italy, researchers needed to get out into the wild to determine its distribution by using trail cameras, trapping, air tubes, and walking transects to assess the population. Only with these tools and information gathered could eradication plans be implemented. Traps and carbon dioxide ($CO_2$) were used to euthanize invasive gray squirrels and Pallas's squirrels to protect the native Eurasian red squirrel population.

Although smaller in weight and size, Pallas's squirrels have a competitive edge by producing three litters per year rather than the usual two litters the other species produce. They also live at higher densities and are more social than native Eurasian reds, said Mazzamuto. Those extra kits born each year gave the invasive Pallas's a leg up on their population and made eradication more difficult. Geography also made full eradication impossible because reaching all the invasive squirrels in wooded areas was difficult. The team, however, experienced success in local eradication. Where those successes occurred, the native Eurasian red squirrels recolonized. Mazzamuto found that encouraging. "The red is pushed at the edges of the optimal habitat, but once you remove that obstacle for them, they go and recolonize the good areas where they once were."

The European Union requires each member country to address invasive species. One of the priorities, said Mazzamuto, was to keep invasive squirrels away from Switzerland and "avoid the transnational move of this invasive squirrel."

Different Italian communities responded differently to the populations of invasive species. One in Liguria chose to sterilize and release the nonnative squirrels in a more isolated park rather than eradicate them. "That connection that the local people had to squirrels" was strong and cohesive. Mazzamuto explained that the people in Varese were more divided because the interloping squirrels are charismatic, eliciting strong feelings among residents, but they also damage infrastructure and orchards.

Unlike the situation in Great Britain, squirrels have yet to be added to menus in Italy, where public opinion is very divided on invasive squirrel eradication and control.

Besides Italy and Great Britain, invasive North American eastern gray squirrels were also introduced into South Africa, Australia, Mexico, and areas of western Canada, Hawaii, and western continental US states. "They become these symbols of globalization and American cultural imperialism, and all this other stuff," said Etienne Benson. All those gray squirrels have become established populations, except in Australia, Hawaii, and Mexico. Within the gray squirrel's natural range, there are five recognized eastern gray squirrel subspecies.

Perhaps the root of the global invasive squirrel dilemma is that squirrel species have been beloved and admired for their small, big-eyed, bushy-tailed appearance and entertaining antics for millennia, so much so that people continue to keep them as pets and then, either purposely or inadvertently, introduce them into the wild, where they often wreak havoc on local native squirrel populations.

And, as Benson reminds us, even as introduced squirrels might be seen as "flattening diversity" in areas where they are threatening native squirrels, their spread and resilience also demonstrates a hopeful and successful Anthropocene story of urban greening and cities viewed as biodiversity hot spots.

The social media story of twenty-three-year-old Venezuelan migrant Yeison, who arrived at the US border in 2023 with his pet squirrel, Niko, exemplifies how strong these human–squirrel bonds become. The pair were inseparable during the journey, with Niko riding along in Yeison's backpack or hat. At the border, the nongovernmental agency Ayudándoles a Triunfar worked with the migrant to secure necessary vaccines for the squirrel. It is likely that Niko will not enter the country with Yeison, but if Niko makes it into the United States, we can only hope he doesn't end up in the wild like many other squirrel pets worldwide.

While invasive gray squirrels are not spreading through Asia yet, another invasive species is causing havoc in Japan. Formosan squirrels (*Callosciurus erythraeus taiwanensis*), native to Taiwan, create headaches for Kamakura residents in eastern Japan. The squirrels are popular with tourists at the city's shrines and temples but have been destructive to the environment. The government has been trapping the squirrels and exterminating them. The rise in the squirrel population in Kamakura might be due to the rising heat caused by climate change. The intense summer heat is believed to have decreased the squirrels' ability to gather mountain nuts, so they show up in urban areas.

Another squirrel crisis exists in the United Arab Emirates, where the five-striped northern palm squirrel (*Funambulus pennantii*), native to western Asia (India, Nepal, and Pakistan), has been introduced by the pet trade, and it is eating fruits and causing damage to crops and cables. Many residents are trapping the interlopers and releasing them in the desert. The advice from the authorities is the same as for preventing most wildlife interactions—remove outside food and water sources and keep entry points to homes sealed.

In the meantime, if nonnative squirrels become as threatening to Italy's local native red species as they are in Britain, perhaps they will start to appear on Italian menus in dishes that traditionally use rabbit, like *coniglio alla cacciatora*, a hunter's-style stew. After all, squirrel cacciatore is a dish prepared by American hunters and foragers in areas where squirrels are commonly eaten.

*Chapter Nine*

# As Dinner

Sciurus albipes Mus. Monac.

*"This food ain't fitten to eat, dogged Georgia rations, Brunswick stew and all. And he's done cooked the squirrel heads in the pur-loo, and that suits a damned Georgia Cracker but it don't suit me."*
—Marjorie Kinnan Rawlings

A political ad for Herbert Hoover once promised Americans "a chicken in every pot." Chickens, made available by factory farming, reduced the need for hunters to strike out into the woods for protein. But that famous phrase has been transformed into "a squirrel in every pot" in Britain's campaign to encourage more squirrel protein in their diet to save their beloved native red squirrel. Perhaps Brits and Italians will take a lesson from squirrel-eating Americans on moving their invasive grays to the table.

Squirrels were hunted and eaten in North America long before Colorado chef Elise Wiggins hunted her first squirrel in Louisiana, long before squirrel cacciatore appeared on plates, beginning with Native peoples. Returning to the story of the Lenape's relationship with squirrels, we know that the relationship between humans and squirrels shifted as squirrels became a food source for humans. Americans hunted for protein long before the urbanization and suburbanization of the country and the industrialization of its food system. Squirrel was just one game meat among many, including possum, raccoon, and turtle, that fit the bill for feeding families nationwide. Eating squirrel meat did, however, gain the reputation for being a "hillbilly-type" food, perhaps in part because of the design of the Kentucky long rifle, created with a small caliber that made it perfect for squirrel hunting. Squirrel hunters didn't need to aim for the squirrel, only for a nearby tree branch, enabling the squirrel to be killed by the concussion of the bullet. Hunger drove the menu.

Appalachian squirrel stew, or Brunswick stew, is a traditional recipe for American Southern cooks. Both Brunswick County, Virginia, and the town of Brunswick, Georgia, claim ownership of the traditional recipe. "The Brunswick Stew. The Originator of It and How the Dish Was First Made" appeared in *The National Republican* on September 16, 1886:

The true history of it is about as follows: During the war of 1812 there was a man named James Matthews, who was a soldier in that war. He was from the red oak neighborhood in Brunswick county, Va. He was also a great squirrel hunter, and it was his way of cooking the squirrel which gained him much popularity and eclat with the ladies.

His mode of cooking the squirrel was quite simple, as follows: After dressing it nicely, the squirrel was set to cooking early in the morning, so that it might be ready for a 2 o'clock dinner. It was kept stewing continually, water being added to supply evaporation, until it was so thoroughly done that the flesh would separate from the bones, which were taken out and the stew seasoned to the taste, not having any vegetable whatever in it. This was the first Brunswick stew, of 1816, and continued to be until 1830–'32, when the tomato had become better known as a most excellent vegetable. About this time a man by the name of Ned Stith (from the same county) conceived the idea of improving "Matthew's stew" by the addition of the tomato, onion, corn, potatoes, middling, fresh butter, and light bread.

Visitors to St. Simons Island, near Brunswick, Georgia, can find a plaque on an old iron pot claiming that the first stew was cooked on July 2, 1898, while Virginia cooks point to a story of the stew being made on a hunting expedition in 1828 by a Virginia state legislator's chef. The recipe also pops up in the 1942 *Cross Creek Cookery* cookbook, authored by Southern writer Marjorie Kinnan Rawlings, who claimed that Queen Victoria favored the dish and it might have originated across the ocean in Braunschweig, Germany—the original Brunswick.

On an episode of MSNBC's *Morning Joe*, 2008 presidential candidate Mike Huckabee discussed growing up in South Carolina. He said, "When I was in college, we used to take a popcorn popper, because that was the only thing they would let us use in the dorm, and we would fry squirrels in a popcorn popper in the dorm room."

Although reputed to be solely a Southern offering, squirrel recipes appeared in Irma Rombauer's *The Joy of Cooking*, a mid-twentieth-century kitchen mainstay. The recipes and a guide to preparing squirrel meat remained in the cookbook from 1943 until 1997.

Regardless of where they originated, squirrels aren't solely on Southern US dining tables. Even into the early 1980s, I saw squirrels hanging from the porch of the local general store in the small Hudson Valley town of Stony Point, just over forty miles from New York City.

Award-winning actor Jennifer Lawrence spoke about being able to skin a squirrel in an interview about the skills she had learned for her role in *Winter's Bone*. "I can shoot quite well, and I didn't think I'd ever be able to say this but, yes, I can skin and cook a squirrel." The Kentucky native told *E! News* in June 2010 that she watched a friend of her brother's demonstrate the skill for her. Television chef Andrew Zimmern got involved in showcasing an array of squirrel dishes, both traditional and new, when he covered a squirrel cook-off in the Ozarks in the second season of his show *Bizarre Foods*. There were even squirrel tamales on the menu.

While the Brunswick stew prepared with an unknown game meat might have been on the plate of Queen Victoria, squirrel meat reached the top plate in the United States around the same time. President James Garfield favored squirrels to the point that the 1887 *White House Cook Book* contains a recipe for squirrel soup, following Philadelphia pepper pot, that requires "three or four good sized squirrels" along with lima beans, tomatoes, corn, butter, and parsley and the use of a coarse colander, "so as to get rid of the squirrels' troublesome little bones."

Garfield wasn't the only president to dine on squirrels. President Grover Cleveland took time out of his schedule to meet up with a farmer friend. This report appeared in the *New York Sun*, reprinted from the *Cincinnati Enquirer*: "Mr. Cleveland brought with him to Oak View five nice fat squirrels as a reward for his hard day's labor, which were served to Mrs. Cleveland and himself for breakfast this morning. Afterward he drove to the Executive Mansion and issued his Thanksgiving proclamation."

In upstate New York, Wade Robertson wrote about the ups and downs of squirrel hunting in the *Olean Times Herald*: "Few things give me the feeling of accomplishment harvesting a limit of squirrels with a rifle does. True, there were some poor decisions, a few missed shots but in the end, all worked out. An iron skillet, butter and squirrel: few things taste better."

"There is absolutely nothing wrong with squirrel meat; our forefathers relied heavily on squirrels as a protein source," wrote Steve Gilliland in 2024 for Kansas's *Hays Post*. He is right.

"One of the proudest moments of my young life was when I took Lady and Tramp, our two fiests [terrier hunting dogs], into a hollow near the house and shot a big fox squirrel. It was the first squirrel I ever bagged while hunting alone and for years I fondly recalled the event whenever I passed that old hickory tree," Professor Terry L. Jones recalled in 2019 of his Louisiana childhood in the 1960s.

Louisiana-born chef Elise Wiggins, who hosted and produced the PBS cooking show *Roots to Ranches*, understands that feeling. Her Italian restaurant, Cattivella, is in Denver, Colorado, where squirrels aren't on every home table but might appear someday. "I grew up in Louisiana, and we hunted anything that moved," she told Laura Shunk in 2017. Buying or selling squirrel meat isn't legal, despite it being readily consumed in the South. Her cooking career has introduced her to many different cuisines, including Italian. Now, her goal is to "get a senator or representative to sponsor me to open a squirrel farm."

Wiggins passionately declares that squirrel meat is the "chicken of the tree." She emphasized to me that unlike the meat of factory-farmed chickens, squirrel meat packs a nutritional punch, with "more protein and nutrients per ounce than the chicken in the grocery store." Wiggins said she would "prefer to eat an animal out of the wild and, as my daddy says, 'do it justice when you dispatch it,' versus a chicken that's in a conveyer belt that hangs upside down and can see each other and then they are dispatched that way.... I'd rather eat something that is better for me that I know died quickly."

"Up here, people say, "What? You eat squirrel?" But think about it: Iberico pig is known because the animals eat acorns. Squirrels are vegetarian and they eat acorns—why would you think the meat doesn't taste good? I know better than to do this in Colorado. But my goal is to become the queen of squirrel meat. I'm sure PETA will be upset at me, but I'm going to wear my squirrel-tail coat and laugh all the way to the bank. Plus, chefs are always looking for something new to use—it's like what wild boar was for a while. And it's delicious!"

Although rabbit gnocchi will appear on Wiggins's menu in 2024, diners will not see squirrels make an appearance, at least for now.

Nowadays, many squirrel hunters in America live in immigrant populations, such as the Hmong, who hunt native squirrels that resemble the squirrels hunted in Southeast Asian mountains. Others are more content knowing where their food originated.

"I realize it's much easier to pick up a chicken at Safeway, but hear me out on this," wrote Jason Salzman, editor of the *Colorado Times Recorder*. "By eating squirrel, you get the added benefit of imagining all the tomatoes, peaches, apples, cherries, Maple trees, and other stuff you're saving from being eaten by the rodents."

The rule of thumb for squirrel hunting and cooking is one squirrel for each person. However, the possibility of infection by mad cow disease or bubonic plague has reduced the number of squirrels on the tables of many Americans, who now opt for venison or chicken on their plates.

I took Chef Wiggins's advice and contacted a hunter to introduce me to the ins and outs of preparing and cooking squirrel. Nancy Heaslip, a retired New York State wildlife biologist, was willing to oblige. She has a view of hunting as a lifestyle rather than a sport, matching Wiggins's outlook. But there are many more Americans who find coexisting with squirrels exasperating. Hunting squirrels becomes not only a sport but also sometimes revenge for chewed wires or pillaged bird feeders. Or, as in Britain, where culling of invasive gray squirrels is encouraged in order to conserve native red squirrels, a matter of species survival.

*Chapter Ten*

# As a Marauder

Plate 113

*"Living off the grid and being kind of an outlaw brings a dangerous reality."*
—Ron Perlman

It's 5:30 a.m. My world is quiet. The only sounds I hear are the heat firing up and my dog eating his breakfast. I'm sitting on my couch, my legs crisscrossed beneath me, my laptop above them, when above me the silence breaks with the sound of scurrying feet and what sounds like a golf ball rolling across the ceiling. Even without X-ray vision, I know it's a squirrel rolling a walnut. Old houses have slivers of openings that enable critters to enter. My old farmhouse is no exception.

"What shall I do? And under the eaves / and through the walls the squirrels have gnawed their ragged entrances—but it is the season / when they need shelter, so what shall I do?" asked Mary Oliver in her poem "Making the House Ready for the Lord."

The simple answer to Oliver's poem and my dilemma is to find those openings and seal them up. But we each have another question to wrestle with: How will we let this disturbance influence our attitude toward or against the squirrel perpetrator?

Discussions about our frustration with squirrels can pop up anywhere, including in the supermarket, at a holiday dinner, or even in a book club discussion. When I mentioned at my book club that I was writing a book about squirrels, one of the women retorted without hesitation, "I hate squirrels." Her direct condemnation of our local squirrels led to a quick discussion on who felt the same and who found them amusing.

Days later, several bird-watchers and I stood in front of a large window overlooking an assortment of seed-filled bird feeders at Hawk Mountain Sanctuary in Pennsylvania. Each feeder had a hefty gray squirrel hanging on, upside down, munching away on the seeds while chipmunks scurried on the ground, eating the fallen seeds—an all-you-can-eat squirrel buffet. The disgust from the bird enthusiast onlookers was palatable.

What a moment to appreciate the irony of this ecosystem's food chain. After all, this was Hawk Mountain, where rodent-eating raptors, including great horned owls and hawks, soar above the rocky cliffs and woodlands and might need a bite before continuing their migration. Those chunky, seed-stuffed squirrels fit the bill. When Rosalie Barrow Edge, founder of the sanctuary and avid birder, founded the Emergency Conservation Committee in 1929, she stressed the need "to protect all species while they were common so that they did not become rare."

The feelings that rise to the surface when we experience squirrels in a park, at a sanctuary, or in our own attic can be strong and often divide us. It's hard for people not to have strong feelings about members of the squirrel family. It isn't only sports and politics that are divisive in America; Americans can also be parsed into two different squirrel camps—those who love them and those who hate them, those who keep squirrels as pets and those who hunt them. For simplicity, let's call them Camp A and Camp B.

Camp A consists of people who feed squirrels outside their windows, finding joy in their playful antics. Camp B consists of people who wage war with the tree squirrels that munch away at their bird feeders, seeing them as pests. Camp A campers join the Squirrels Are Awesome social media group, while Camp B campers join the Squirrel Haters of America group. Then there are those independent campers who waffle between both camps, appreciating squirrels' beauty but considering the need to protect their own property. Many fall into this camp. Regardless of group identity, squirrels can elicit strong feelings from anyone at any time, including at book club meetings or perhaps your next Thanksgiving dinner.

"People with bird feeders bring up their frustrations to me all the time," Noah Perlut told *Washington Post* journalist Kate Morgan. "I'm always curious why they don't like the squirrels. Aren't you feeding wildlife because you want to see them? Why is it you only want to see birds, rather than birds and squirrels together?"

Wildlife coexists quite well without us, but we often need help when we interject ourselves, our feeders, and such into the mix. Even those on the live-and-let-live team, Camp A, who enjoy seeing squirrels at their feeders and munching on their Halloween pumpkins, can understand and sympathize with those frustrated individuals in Camp B, primarily when a squirrel scrambles across a ceiling or causes a power outage.

We can sympathize with backyard birders awaiting the sublime flutter of wings at their feeder. However, finding a gobbling squirrel swinging from it like some poor parody of Miley Cyrus's *Wrecking Ball* video can be more than disappointing.

Terry Rich, a contributor to the *Idaho Press*, wrote about his ongoing personal squirrel skirmishes with fox squirrels in Boise: "We have a western screech-owl box in our yard, in which we want western screech-owls. Keeping bushy-tailed tree-rats out of our owl box is a nearly full-time job. Color me grumpy."

The strong words he writes might be about those same fox squirrels I spotted in the trees in Julia Davis Park. *Tree-rats* sounds a bit harsh and grumpy but perfectly describes his feelings. He's definitely in Camp B. It isn't that Rich doesn't like the squirrels that welcomed me to his home city in Idaho, but those same little park squirrels are getting in the way of attracting the presumably more charismatic wild owls that he desires. Who doesn't love owls? Rich's story reflects something more profound—our conditional love of wildlife. I will love you and accept you if you stay out of my owl nest boxes and don't eat my birdseed or dig up my flower bulbs and make a mess. I will love you if you don't behave like a wild animal.

Right or wrong, sometimes we love a species until it causes us disruption. But we must accept that squirrels are very good at being squirrels. Disruption is their middle name. And while we're talking about disruption, we must come to grips with the fact that humans have cornered the market. We've put up the feeders. We've clear-cut their forests. Don't throw stones from glass houses.

The outdoor bird feeder market in the United States was valued at $216.8 million in 2025. It is expected to increase to $309.15 million by 2031. Keeping bird feeders squirrel free has become an industry standard for people searching for squirrel-resistant, chew-proof feeders that keep squirrels out and birds in. Birders can find feeders with names like Squirrel Buster among their options, as well as feeders that include "squirrel baffles." Birders are encouraged to use strong scents, such as garlic, peppermint, and pepper, on their feeder poles to deter climbing squirrels from eating the seeds purchased for wild birds. But squirrels see a buffet placed out for them and will continue to attempt to reach it despite the difficulties. That demonstrates just how strong their drive is to obtain food to survive.

That drive is just one of the reasons it is illegal to feed squirrels in places such as Ontario, Canada. Feeding squirrels and other wildlife in your backyard might seem friendly, but it upsets the food chain and can lead to unintended encounters. As prey, squirrels can attract predators as they move closer to humans. In addition, squirrels can gnaw through wood window frames for more food.

However, the squirrels' drive to obtain food isn't limited to birdseed or peanuts. Their omnivore orientation also drives them to a marauding behavior that extends beyond bird feeders. And birders aren't alone in wanting to discourage squirrels.

Bill Carver, a campground host in state parks and national forests, always deals with the sound of nibbling squirrel teeth on the top of his recreational vehicle roof. But it isn't just the noise that grates on him; it's the chewing of his insulation and the cover of his grill, as well as the fear that the squirrels will eventually chew his electrical wiring or sewer hoses. To lighten the mood and foster a sense of community, he created the satiric Squirrel Haters of America group on Facebook. Here, he and others frequently poke fun at their squirrel frustration with humorous squirrel jokes and cartoons, such as one depicting a squirrel telling his human therapist, "When I learned, 'You are what you eat' I realized I was nuts." A brief questionnaire to join the group makes it clear that the group is designed to bring humor to our complex relationship with squirrels, providing a much-needed chuckle in the face of squirrel-induced exasperation.

Carver admits that *hate* is a strong word and that he actually enjoys watching the squirrels' antics. Carver, like much of the population, has a love-hate relationship with his local squirrels. His social media group unites everyone in their shared exasperation with amusement. Carver's Facebook account is not the only one—there are many focused on squirrels. Across the social media world, squirrel fans will find fewer jokes and more photos of adorable squirrels, even some dressed up in clothes in another Facebook group with thousands of members called

Squirrels Are Awesome and another, broader group that can unite everyone titled All Things Squirrel.

Gardeners, another group vexed with squirrel marauders, have their concerns about keeping squirrels away from their well-tended gardens. The recommended use of coffee grounds, pepper sprays on the soil, and sprays on plants to deter squirrels from uprooting plants and stealing bulbs is a testament to their collective struggle. For these nature lovers, preventing natural squirrel behavior is key to enjoying their selection of nature, whether it's a feeder hanging from a tree or a well-tended garden.

While we might resent the squirrels in our gardens and at our feeders, it's crucial to remember that these backyard "terrorists" are also exhibiting much darker behavior in their quest for food. As omnivores, squirrels spend a considerable amount of time pillaging the nests of wild birds. This behavior is not often witnessed, but it's a stark reminder of the complex nature of these seemingly innocent, adorable creatures.

"Was it the gray squirrels? The red squirrels?" wrote Diane Ackerman in *Cultivating Delight* after the gruesome discovery of a devastated wren nest in her garden. "Either could have been the culprit. One is not supposed to take sides, but I feel the hard punch in the stomach that is loss. All the work the wrens put into raising their young—then to helplessly watch them mutilated and devoured! What horror."

Ackerman's grief is shared by anyone who witnesses the cruelty of nature unfolding around them, including me. On more than one occasion, I've watched frantic birds attempting to keep their nests safe from marauding gray squirrels. Most recently, it was the Carolina wrens building a nest in the nest box mounted to a maple, trying desperately to ward off a diligent gray that climbed the bark toward it. Over and over again they flew at the gray squirrel.

A friend witnessed a hawk flying off with a baby woodchuck that spring. When I shared with her the story of another friend seeing a hawk fly off with a squirrel, she added, "See, we have a lot of squirrels, and while I'd be sad . . . I am not attached to the squirrels. The hawk babies can have all the squirrels and chipmunks they want from my yard. Just leave my pups alone!"

Everyone has their own bias about their local squirrel species. Is the impression that squirrels are pervasive and disposable reflected in my friend's response? And would it change if her local squirrel species was uncommon or endangered?

While appreciating the natural world's beauty, it is imperative to accept the parts we might perceive as ugly. As the saying goes, without shadow, there is no sunlight. The darkness of the predator–prey relationship is never easy to witness. We love watching recordings of lions in the wild until they pounce and devour a zebra foal.

"I know this is the way of nature, kill and be killed, and that the squirrels and raccoons have young to feed," wrote Ackerman.

Ackerman is correct. All creatures have a purpose and the right to exist in nature's endless struggle. One creature is not above another. The fisher cat catching the squirrel is as vital to its ecosystem as the squirrel preying on a vole or young birds.

We can't forget the role we also play. We provide feeders and nest boxes to feed our spirit and assist the wild creatures we've displaced with our buildings, nonnative gardens, and presence. However, our need to bring nature to us sometimes creates more opportunities to witness its rawness.

"Both red and gray squirrels will eat birds' eggs and nestlings, Ackerman wrote, but especially red squirrels, and they're small enough to ease in and out of the box." Squirrels are omnivores and, as gruesome as it sounds, will raid bird nests to eat eggs or chicks. Those bird eggs and nestlings provide squirrels with a crucial boost of calcium, protein, and essential vitamins and minerals they may not get as quickly from tree nuts. However, squirrels aren't the only ones preying on bird eggs and nestlings. Many birds, such as blue jays and hawks, also prey on eggs and nestlings. It is sometimes a bird-eat-bird world out there.

While squirrels are targeted as bird nest predators, they also serve as bird protectors by providing an early warning system that birds often rely on when foraging on the ground. Birds and squirrels are vulnerable to other predators from above and on the ground, such as predatory birds and prowling cats. That squirrel–bird relationship can be complex because, in addition, both squirrels and birds sometimes use each other's nests.

When squirrels aren't busy warning wildlife or marauding feeders, gardens, or bird nests, they can also get into trouble with humans by chewing on things other than food, such as car wires or house siding. It's their nature, as card-carrying members of the order Rodentia. "The reason they are chewing on things is because they have incisors—front teeth—that are always growing," explained wildlife biologist Robert McCleery of the University of Florida's Institute of Food and Agricultural Sciences. "If they don't chew on something, their teeth are going to grow into their lower jaw and skull." To avoid that, squirrel family members, including marmots, will chew on anything that helps wear down their teeth.

Car wires are just the start. Have you ever heard squirrels scurrying above your ceiling, as I have, and cringed at the thought of them running through your attic or scrambling down your chimney?

Connecticut's Department of Energy & Environmental Protection posts a page online titled "Squirrel Nuisance Problems" with advice to homeowners. "A squirrel trapped in a chimney should not be removed through the fireplace area because it might escape into the room. Instead, lower a heavy rope down the chimney to provide the means for the animal to climb out. Drop the other end of the rope to the ground to avoid another trip to the roof to retrieve it after the squirrel has left."

While this makes sense, there might be more practical solutions than getting on your roof to drop a rope down. I couldn't have imagined climbing on our tile roof to drop a rope down the fieldstone fireplace chimney of our early 1900s Hudson Valley home when a squirrel became trapped in our family chimney. It caused such chaos as it ran about the house covered in chimney soot. Ridding your home from squirrels is a job for professionals. But there are many ways to deter squirrels from entering your home in the first place. Neighbors may suggest cutting down the trees around your house. While that is an option, it might not be one you are willing to live with. Using humane traps to catch and release the squirrels is another option. Sealing up holes where they can enter is the best solution.

Regardless of how humans try to fend off marauding squirrels, squirrels still can cause power outages and other problems that wreak havoc. A satirical website named *Cyber Squirrel 1*, developed by Cris

Thomas, documented power outages attributed to squirrels in public records between 1987 and 2019. As of 2019, he had documented 1,252 outages worldwide caused by squirrels.

Not everyone allows the disruptions to derail them. Some individuals and towns embrace the chaos squirrels instigate, embracing the attitude "If you can't beat 'em, join 'em." The way Piedmont, Alabama, residents celebrate New Year's Eve is a perfect example. The Southern city doesn't drop a ball on New Year's Eve, as done in New York City, or a peach, as in Georgia; instead, the city "drops a squirrel" each New Year's Eve to mark the start of the new year and poke fun at the cause for power outages throughout the year. Each year, a mechanized Sparky the Squirrel cutout runs up a faux telephone pole and ceremoniously sets off a transformer, creating sparks at midnight on December 31.

However, the community of Piedmont isn't alone in experiencing squirrel-induced power outages. In 1987, a squirrel stepped onto a power line in New York City, met its demise, and caused an outage that shut down the National Association of Securities Dealers Automatic Quotation System (Nasdaq) for eighty-two minutes. Another squirrel caused another Nasdaq disruption in 1994. And there have been more power outages. Almost every week, a reported outage in America is attributed to squirrel activity. During one recent January week, squirrels caused power outages in at least three locations: a shopping mall in Santa Rosa, California; in Dixon, Kentucky; and in Flagstaff, Arizona. A squirrel caused a two-hour power outage in 2018 for twelve thousand residents of upstate New York. Among other outages, one in 2022 caused a blackout for over ten thousand Virginians. While these significant squirrel-induced outages didn't lead to memorializing those squirrels, another university campus had its own Sparky moment.

On Utica University's campus is a memorial dedicated to its own Sparky by the Class of 2018 with a small squirrel statue inscribed, "The squirrel that searched for light even during the darkest of times." Sparky the Squirrel nibbled on a live wire, causing a campus-wide power outage right before the 2015 conference men's basketball championship game. Each October, the campus holds a celebration named Squirrel Day in its honor, complete with warm cider, games, and emotionally charged eulogies to the original Sparky.

As we've seen, some colleges revel in their squirrel populations. As campus celebrities on Bryan College's Tennessee campus, squirrels were called out in a satire piece by student Rob Speed, a college journalism major and head of satire for the college's publication, *The Triangle*. Speed put forth a conspiracy theory in the Bryan Triangle that alleged the campus squirrels were conspiring with skunks and other local woodland wildlife to one day overthrow the campus and "harm us all." His suggestion to combat this threat was to let out a scream at every squirrel sighting: "Don't be passive to the usurpation of our campus," he implored. "Together, we can ward off the invasive revolt of small woodland creatures—one scream at a time." A different approach to ubiquitous campus squirrel populations invokes amusement and frustration.

As much as Speed's posting provides a laugh, many other memorials and celebrations recognize the usual squirrel shenanigans that can often cause power outages and other damage. Although these memorials and celebrations are not frequent, squirrel damage can be troubling and costly. The internet overflows with ways to eradicate pesky squirrels, ranging from humane and ethical approaches such as pepper spray repellents, decoys, live cage traps, and elimination of entry holes in homes to the more radical and environmentally harmful solution of poisoning them or using lethal traps.

Returning to that squirrel I observed in Maine at the Newagen Seaside Inn, where biologist and author Rachel Carson wrote, I wondered whether that busy squirrel descended from one Carson may have observed from the same spot. She wrote about the poisoning of squirrels in the pages of *Silent Spring*, a tragic reminder that not all squirrel stories are as warm and fuzzy as the one I recounted. Humans have a tumultuous history with these industrious rodents. The squirrels she included were found poisoned by pesticides in Michigan: "Among the mammals ground squirrels were virtually annihilated; their bodies were found in attitudes characteristic of violent death by poisoning.... The fox squirrel had been a relatively common animal in the town; after the spraying was gone."

Carson didn't mince words in her description of one poisoned squirrel, which "'exhibited a characteristic attitude in death. The back was bowed, and the forelegs with the toes of the feet tightly clenched were drawn close to the thorax. . . . The head and neck were outstretched and the mouth often contained dirt, suggesting that the dying animal had been biting at the ground.' By acquiescing in an act that can cause such suffering to a living creature, who among us is not diminished as a human being?"

Our collective observations run the gamut from the ones told orally in traditional stories to the gentle one in this bucolic seaside setting, to the ones long ago in Michigan, poisoned before DDT was banned. Our stories of squirrels, of that "little opera," are tales of squirrels being beloved and bullied, entertaining and enraging. Praised and poisoned, as the ones described by Carson or purposely.

When people use a rodenticide, an anticoagulant mixed with other chemicals, the poison causes the animal's blood vessels to weaken, causing internal hemorrhaging. The death is slow and painful, with squirrels and other small rodents dying over many days. If they are preyed upon during those days, secondary poisoning can occur throughout the ecosystem. Hawks, owls, foxes, bald eagles, endangered California condors, mountain lions, and even domestic cats and dogs can also suffer the effects of the poison when they catch and eat a poisoned squirrel. These subsequent deaths are contributing to the widespread decline of Europe's birds.

A 2021 scientific investigation by scientists of the Leibniz Institute for Zoo and Wildlife Research (Leibniz-IZW), the Julius Kühn-Institute, and the German Environment Agency found these poisonous substances "widely found in liver tissues of birds of prey from Germany." These birds included northern goshawks, red kites, and white-tailed sea eagles. "We found rodenticide residues in liver tissues of more than 80 percent of the northern goshawks and red kites which we examined," said Alexander Badry of Leibniz-IZW, the lead author of the article, published in *Environmental Research*. Not all homeowners are to blame for these deaths in Germany. Anticoagulant rodenticides are often used on large German plantations to protect trees and to control rodents in the sewage systems and canals of towns and cities.

Of course, Germany isn't alone. In 2024, a beloved owl perished in New York City. Flaco was a Eurasian eagle owl who had lived in the Central Park Zoo before the wires of his enclosure were cut, enabling him to escape into the wilds of the city park. He became quite a celebrity and was photographed daily in trees and atop buildings. Sadly, Flaco was found dead in February 2024. Initial reports indicated trauma, which upheld the speculation that he had flown into a building, but a full necropsy report was planned to find out whether there was any potential exposure to rodenticides or other toxins that might have contributed to the famous bird's death.

Barry, another celebrated owl that called Central Park home, died in 2021 from colliding with a car, but tests later showed that the barred owl had high levels of rat poison in her system. Poisoned birds, like Barry and Flaco, can suffer from a lack of coordination and a weakness that can facilitate collisions with cars and buildings. Rodenticides are applied by the New York City Department of Health and Mental Hygiene each year. NYC Bird Alliance noted that rat poisons were detected in 84 percent of the city's dead birds in research conducted by the New York State Department of Environmental Conservation. A total of 75,749 pounds of toxic rodenticides were applied in New York City by city agencies in 2022. Did Barry and Flaco digest too many poisoned squirrels? These deaths are a stark reminder of the preventable nature of these tragedies.

No one wants to think about poisoned squirrels during a beautiful fall day in Maine. I wanted to believe that the little red I observed was, in fact, one in a long line that had brought joy to the guests at Newagen and Rachel Carson as it went about the business of being a squirrel.

People can address the issues they have with nuisance wildlife without applying poisons, but some ways of driving squirrels from a home or feeder might be outside their control, whether or not they mind their presence. Humans can also be proactive by adding a squirrel feeder to the mix of feeders and nest boxes on their property, which might be the answer to squirrel woes. This proactive approach can lead to a hopeful future of coexistence with wildlife.

Although wildlife experts advise against using squirrel feeders because it contributes to making wildlife dependent on humans and inhibits their ability to search for their wild food, many such feeders on the market offer wildlife watchers a way to enjoy local squirrels instead of suffering with exasperation, thinking that if the squirrels use their squirrel feeders, they will not be attracted to bird feeders. Of course, this assumes that the squirrels will know which feeder is intended for them. If you do offer your squirrels a tasty treat, at least make it something they would seek out in their own natural environment, like tree nuts in the shell. Providing local tree nuts is the best option, whether they are walnuts, hickory nuts, or white acorns. Avoid peanuts and other nuts that are not something that they usually eat in the wild and that offer less nutrition.

The best solution to encourage all wildlife to visit your property is to leave it as natural as possible, avoiding raking fall leaves to provide shelter for wintering bees, mowing later in spring, and planting native plants that enable squirrels to find shelter and food throughout the seasons. Doing this in coordination with using natural ways to discourage frustration will encourage coexistence, keeping you and your home in harmony for yourself and the local wildlife sharing the habitat with you. History has demonstrated that squirrels might just leave on their own when nature, for one reason or another, has become an inhospitable host.

*Chapter Eleven*

# On the Move

> "The behavior of squirrels during September seemed unusual according to many observers. Squirrels appeared in areas where they were not usually seen, and one man commented on the 'dancing' of squirrels on the highway."
> —Vagn Flyger

The year 1968 was a year that shattered America. It was the year when the lid of the simmering, bubbling pot of the civil rights movement, youth culture, the Cold War, and the Vietnam War exploded. The year began with a march of some five thousand women in Washington, DC, to protest the Vietnam War. The same month, the first televised NCAA basketball game appeared in prime time, North Korea seized a US ship, and North Vietnamese communists launched the Tet Offensive. That spring, New York Senator Robert F. Kennedy entered the race for the Democratic presidential nomination, and Martin Luther King Jr. was fatally shot in Memphis, Tennessee. Soon after, students took over five buildings on Columbia University's campus, calling for the university to cut its ties to military research, leading to the arrests of seven hundred people, and over one hundred people were injured. The same week, *Hair* opened on Broadway. Peace talks began. Andy Warhol was shot and wounded. President Lyndon B. Johnson signed the Treaty on the Non-Proliferation of Nuclear Weapons. The Soviet Union invaded Czechoslovakia. The song "Hey Jude" was released in the United States and became the longest-running song to hit Number 1 on the *Billboard* Hot 100.

Then autumn arrived, and with it came the Great Squirrel Migration. Amid the news of the Apollo 7 mission and the Olympic Games in Mexico City, people in the eastern United States experienced an event of nature unlike anything they'd witnessed. Gray squirrels were moving by the thousands out of woodlands and parks, crossing mountains, rivers, and roads.

Squirrels are customarily seen as roadkill on America's roads, partly because they can change direction on a dime. They are the most commonly killed animals on America's roads. That ability is a beneficial skill for avoiding predators but not cars.

Even I am not immune to having experienced striking a squirrel with my car. The gray squirrel seemed to be on a suicide mission as it ran about the road and straight into my wheel. But, in fact, it was not. It was doing what squirrels do to escape a predator. They scramble. Unfortunately, my car was not a confused coyote, and it struck the tiny creature, killing it instantly.

An estimated 41 million squirrels are struck and killed each year, according to *World Atlas*. But the autumn of 1968 presented an unusually high number of squirrels struck on the roads. The New York State Department of Environmental Conservation collected 122 squirrel carcasses on the roads near Albany. Dozens reported to the agency had drowned in the nearby Hudson River. In the lower Hudson Valley, the same migration was occurring. *The Journal News* in Westchester County ran an article titled "Squirrels Invading Lower Hudson Valley" on September 28.

Before the Great Migration of 1968, the last reported occurrence was in 1933, a thirty-five-year spread. What made the 1968 event truly remarkable was the coordinated scientific study that accompanied it, a rare occurrence in the world of animal migrations.

A North Carolina hunting forum member wrote in 2018 about his memory of that year: "There were squirrel everywhere. I have never eaten as many squirrels as I did that year. I never knew what caused it, nor have I seen anything like it since. Any boy with a shooting iron could get their limit of bushytails in just a few minutes here in Macon County."

The *Asheville Citizen Tribune* reported on the migration with a story on September 17 titled "Starvation, Cars, Killing Squirrels by the Thousands." *The New York Times* published another story, "Squirrels Starving in Smokies Area," on September 22.

What was going on? The squirrels experienced a bountiful mast year for acorns and chestnuts in 1967, which created a boon for the squirrel population. A mast year, when species of trees and shrubs produce a bumper crop of fruits or nuts, occurs every few years, on average three to five. This pattern creates a "predator satiation" situation. The year following a mast year is usually a bust for the increased population. The nut-bearing trees produced an insufficient amount of nuts for the squirrels. Mast years affect animals such as squirrels, mice, and jays that feed on acorns. This was the case in 1968. The bushy-tails had no choice but to search for more fruitful trees elsewhere. So, in September 1968, the squirrels left the Northeast in droves for other parts of America.

Reports on the migration continued, and information was released from each state and presented on November 4, 1968, by the Smithsonian Institution's Center for Short-Lived Phenomena. A paper was presented at the Northeast Fish and Wildlife Conference in White Sulphur Springs, West Virginia, the following year by Maryland's Vagn Flyger, a wildlife biologist and renowned authority on squirrels, who noted that the behavior of the migrating squirrels was also unusual:

> The behavior of squirrels during September seemed unusual according to many observers. Squirrels appeared in areas where they were not usually seen, and one man commented on the "dancing" of squirrels on the highway. As cars approached, squirrels on the road would sometimes jump into the air, zigzag back and forth, and behave erratically. I also observed this. One man explained this by saying that the migrating squirrels had traveled so far that the soles of their feet were very thin and therefore they felt the vibrations of approaching cars, and the tickling sensation made them jump into the air. A more likely explanation is that emigrating squirrels caught out in the middle of the road were unfamiliar with the area and, not knowing which way to run, behaved erratically because they were terrified and confused.

At the end of his report, Flyger concluded that squirrel migrations are "poorly understood and should be investigated in greater detail to determine their causes."

An article by Daniel Smiley in *The Chirp*, the bulletin of the John Burroughs Natural History Society, published at Mohonk Lake in New York's Shawangunk Ridge in November 1968, offered another opinion. Smiley wrote that a "recent cruise on the Hudson [River] had revealed 30 dead squirrels in 1-1/2 miles. However, the additional comments from my informants were to the effect that although acorns and nuts are not as abundant as in some years, they are *not scarce*; the squirrels are *not raiding* corn fields and other cultivated crops; and those found dead are in *good condition* (not starving)." This article offered the opinion that it wasn't a lack of food that drove them on their migration.

While this was unusual, it was not the first time this migratory behavior had been observed in the Hudson Valley or elsewhere. The *Chirp* article cited records of the Research and Records Committee at Mohonk that showed that the "previous gray squirrel population peak and emigration was in the fall of 1961" and recorded other moves in 1953 and 1954, both at seven- or eight-year intervals. Other parts of the country had seen their share of migrations too.

Migrations had been noted throughout the 1800s on an average of every five and one-quarter years and in the 1900s roughly every seventeen years. On September 11, 1803, twelve days after Meriwether Lewis left Pittsburgh, Pennsylvania, on his voyage down the Ohio River, he put pen to paper to write a description of eastern gray squirrels swimming in the river:

> Observed a number of squirrels swiming the Ohio and universally passing from the W. to the East shore they appear to be making to the south; perhaps it may be mast or food which they are in serch of but I should rather suppose that it is climate which is their object as I find no difference in the quantity of mast on both sides of this river it being abundant on both except the beach nut which appears extremely scarce this season, the walnuts and Hickory nuts the usual food of the squirrell appears in great abundance on either side of the river—I made my dog take as many each day as I had occation for, they wer fat and I thought them when fryed a pleasant food—many of these squirrils wer black, they swim very light on the water and make pretty good speed.

Meriwether Lewis and William Clark continued to see gray squirrels along their route and eventually observed them as high up the Missouri River as the mouth of the Little Sioux River, estimated to be about 733 miles from the Mississippi. Later, in 1806, Lewis wrote about the Columbian ground squirrels (*Spermophilus columbianus*, now referred to as *Urocitellus columbianus*).

One migration, in 1833, became legendary and the topic of a classic 1994 episode of *Across Indiana*. Millions of squirrels marched into Salem, Indiana, from the north that year. According to reports in the local papers, the squirrels were seen passing right through houses and farms. Zoologist Julian Duval of the Indiana Zoo was skeptical of the story, asserting that he didn't "believe there's any good scientific record for any migratory habits of squirrels; they would tend as a group not to be migratory animals."

While some reports have the squirrels coming from the north, out of the woodlands, and into the farmlands, others claim they invaded from Kentucky to the south. Reports suggest that the squirrels swam across the Ohio River by the thousands, suggesting that there were so many that a person could have walked across the water on top of them. Even John James Audubon noted the swimming squirrels. Like hungry locusts, the squirrels were said to have eaten everything.

When a devastating mid-spring cold snap in Texas in 1857 killed off crops and vegetation across the state, squirrels took off for actual greener pastures. One recorded account from a young man named Henry Garrison Askew described thousands of squirrels crossing the road near Dallas; some were claimed to have scurried right over the horses pulling his carriage.

While a lack of food or habitat contributes to shifts in the populations of many species, other theories about migration include an increase in agitating fleas, competition from aggressive red squirrels, the closing of tree cavities with concrete, an overcrowding population, and even sunspots.

In the end, the answer was even more complex. During the 1800s, gray squirrels were constantly pushed into smaller areas where thousands of squirrels already had homes, creating unnatural competition. But it wasn't just the squirrels that were being stressed to relocate. Hawks, foxes, and wolves were also being pushed into these areas, which increased anxiety among the squirrel population. So, as winter approached, squirrels trying feverishly to create stores of food were finding their situation untenable.

In his *AtCharlie Chronicles* in 2023, Steven Veigel wrote his thoughts about the migration: "It didn't happen all at once, but squirrels naturally started moving looking for somewhere they could spread out, have a home and do what they had been doing for millions of years. But there were just too many." Perhaps Veigel was right. Maybe the squirrels found that they didn't even have enough time to bury an evening meal amid increased natural predators and human hunters firing guns. No wonder they were scrambling to get out of there.

He continued: "A hundred squirrels trying to find room to spread out with food to eat. They meet another hundred. Traveling through already highly populated areas. Then there's a thousand. And then 30,000. Swimming rivers, eating crops to survive on their journey. Just trying to get to open land. Then the swarm spreads out and the 'migration' is over. It's not that difficult or mysterious. People just don't include themselves or predators in the study. How they tend to create the thing they fear."

As demonstrated by Veigel, observing and recording squirrel behavior isn't limited to scientists. Blogger, photographer, and naturalist Michael Bateman wrote about the decrease in his local New Jersey population of squirrels and chipmunks in 2021. Following a mast year in 2019, a significant increase in rodents occurred in 2020, just as everyone in his area was setting up new gardens during the COVID-19 pandemic and found their produce pounced upon by nibbling rodents. But in 2021, that surge of squirrels led to fewer food sources and a population collapse. Many wildlife enthusiasts in his area witnessed unusually high squirrel roadkill on area roads. There was even a report of a squirrel swimming in a lake. Squirrels were on the move.

Another mast year occurred in 2023. As I write, it's 2024, and we have yet to see whether squirrels took to burying the abundance and there will be a resulting increase in oak seedlings and a crash in the squirrel population in the near future. But there was another finding. Researchers with the Northeast Climate Adaptation Science Center found that these masting events affect red squirrel populations, and an increase in red squirrel numbers leads to a decrease in songbird nest survival rates as the reds expand their range into higher-elevation spruce-fir dominated forests and prey upon the migratory songbird species. The finding will inform northern forest management and songbird conservation in the Northeast.

Could we see more migrations as squirrel habitats become uninhabitable during this period of human-induced rapid environmental change (HIREC) and the world continues to lose biodiversity in great numbers? "We've seen climate change pushing, so it's not just human involvement in introducing these species; we're creating a world where a lot of species have to move," said science historian Etienne Benson. We see that "when they move, they either possibly replace something else or compete with something else."

Will a natural pattern in squirrel migrations after mast years proceed? If we consider this pattern an example of just one such pattern in the natural world, examining it can help us learn more about the changing rhythms of nature in this changing world.

A groundbreaking 2023 study published in *Biological Reviews* led by PhD student Catherine Finn and Daniel Pincheira-Donoso from the School of Biological Sciences at Queen's University Belfast and Florencia Grattarola from the Czech University of Life Sciences, Prague, examined the population densities of over seventy thousand animal species across the planet. This comprehensive research, conducted on a global scale, led to the finding that almost half (48 percent) of Earth's species are currently undergoing population declines. The researchers also noted the disturbing finding that 33 percent of species currently considered safe by the International Union for Conservation of Nature and Natural Resources (IUCN) are declining toward risk of extinction. In addition, less than 3 percent of species were found to be experiencing population increases: "Whether species and their populations can survive the Anthropocene defaunation will depend on their intrinsic traits, their adaptive potential, and also the research and management we dedicate towards preventing their disappearance." In addition, the United Nations estimates that one million species are threatened with extinction. Many scientists believe that number is actually higher.

While it is estimated that there are between two hundred million and three hundred million individual squirrels in the world, population data on their overall health depends on each particular species. *Ubiquitous* doesn't define the entire squirrel genus. Populations of each species vary from location to location and are dependent on food and habitat security. They can drop by as much as 90 percent in an area as other regions see an increase. And the studies are scarce.

An example of this can be seen in the western gray squirrel (*Sciurus griseus*) population, which has declined dramatically since the late nineteenth century in Washington state and is more populous in other regions. Its isolated population is suffering from habitat loss and degradation, wildfires, highway mortality, and disease. Many of these threats are exacerbated by climate change. "Their populations are going down in some regions as their natural habitats are destroyed. Despite their abundance, squirrel numbers are falling all over, and if continued, they may not survive the next century."

So, how do we move forward when the responsibility of protecting all wildlife species, including squirrels, in the United States currently falls on state fish and wildlife agencies, which are predominantly perceived as game wardens with limited funding?

As an example, an Associated Press analysis of 2020 data found that fish received 67 percent of the $1.2 billion per year spent on recovering endangered and threatened species, the majority of which was designated for "several dozen salmon and steelhead fish populations in California, Oregon and Washington. Mammals were a distant second with 7% of spending, and birds had about 5%. Insects received just 0.5% of the money and plants about 2%. Not included in those percentages is money divided among multiple species."

A former director of the US Fish and Wildlife Service, Jamie Rappaport Clark, said that debating the allocation of scarce resources for rescuing endangered species is a distraction. "The issue is not where the money is spent," said Clark, who is now president of Defenders of Wildlife. "The issue is that there isn't nearly enough of it."

Legislation brought us the Endangered Species Act of 1973, and legislation can answer the funding problem in this challenging time of biodiversity loss. A bill originating in the United States Senate, S. 2372, Recovering America's Wildlife Act of 2022, also known as RAWA, would direct nearly $1.4 billion annually to state and tribal wildlife agencies to protect species of greatest conservation need, often overlooked species. The original bipartisan congressional bill (H.R. 2773) was introduced by Representative Debbie Dingell (D-MI) in 2021. "We are in the midst of an unprecedented biodiversity crisis. The Recovering America's Wildlife Act would be a landmark change in the way we fund fish and wildlife conservation. Without action, the list of federally threatened and endangered species will grow from nearly 1,600 species today to thousands more in the future," said Dingell.

The bill passed the House (231–190) in June 2022, went on to the Senate as S.2372 in 2021, and was referred to committees of jurisdiction. In March 2023, it was reintroduced in the Senate by Senator Martin Heinrich (D-NM) as bipartisan bill S.1149 and again referred to committees of jurisdiction. The bill would amend the outdated 1937 Pittman-Robertson Wildlife Restoration Act, which supports wildlife species targeted by sportsmen and game hunters. It would provide funding to recover endangered species and direct funds to tribal nations to support environmental conservation on 140 million acres of land. It would give more funding to less popular and charismatic endangered species. The funds would be derived from fees and penalties.

"RAWA represents a strong commitment to addressing this crisis using innovative, on-the-ground collaboration that will protect our nation's environmental heritage for years to come. Grateful to advance this bill alongside my partner in this effort Congressman Jeff Fortenberry, Chairman Grijalva, and the broad, bipartisan coalition that has fought for this legislation's purpose," said Dingell before the House vote.

"There are at least 12,000 different species in need of proactive efforts to prevent them from becoming endangered. There are all these needs out there that the current funding structure is not capable of tackling," said Sara Parker Pauley, director of the Missouri Department of Conservation and former president of the Association of Fish and Wildlife Agencies, a Washington, DC–based coordination and advocacy association for the state agencies.

RAWA has bipartisan support as well as backing from environmental organizations including the National Audubon Society, The Nature Conservancy, and the World Wildlife Fund (WWF).

Although RAWA might promote more studies of squirrel populations throughout the United States, its future is uncertain. Although

President Donald Trump implemented 145 actions targeted at dismantling environmental protections, the Recovering America's Wildlife Act still remains in play. Its future is unclear, but it would be a great tool and would facilitate survival for some squirrel species.

## Chapter Twelve

# By the Numbers

*"To be counted in the census is to be both seen and supported."*
—Alex Wagner

Walking with a friend in New York City's Central Park, off Fifth Avenue, I might have been in any urban park. On this hot, sunny day, people poured out of their offices and apartments to enjoy the green space. There were joggers, book readers, birders, tourists with cameras, dog walkers, and people enjoying lunch.

But, unlike many other urban parks, this one is vast and found in what the *Encyclopedia Britannica* describes as the "largest and most influential American metropolis." The 843-acre park hosts over 42 million visitors each year. Central Park's resident squirrels see them all. From their sycamore branches, rocky perches, and grassy hills, the park squirrels spot babies pushed in strollers, gatherings of students wearing matching tour T-shirts, couples holding hands, and teens with headphones.

While we know how many humans walk the park's paths, we are still determining the exact number of squirrels that scurry across them. On my lazy, hot summer afternoon stroll, I spotted only two on a mere fraction of the park's trails. But I discern that there are many, many more. How do I know?

"No census has ever been taken of the squirrels of Central Park, but, if it were possible to count them, their number would probably fall much below common expectation. Some people fancy that there must be thousands, but there are in fact probably not more than two to three hundred, if so many; they might not really exceed 150." The revelation of this low estimate, which wasn't written yesterday, is quite surprising. It appeared in *The New York Sun* on October 21, 1900, when a census didn't exist. It would be more than a century before a census was undertaken to achieve a more accurate view of the park's squirrel population.

The 2018 Central Park Squirrel Census, an event supported by the Explorer's Club, the Central Park Conservancy, New York University, NYC Parks, and many volunteers, with a social media shout-out from the mayor's office, attempted a more realistic estimate of the squirrel population. Those census takers, dubbed Squirrel Sighters, spent much more time in the park than most visitors, including me.

Indeed, the three hundred volunteers who diligently tracked down the bushy-tailed residents of the park spent eleven days on their census count. Their dedication was unwavering. For the task ahead, the park was divided into hectares, each roughly the size of a squirrel's home territory (a hectare equals 2.47 acres). These hectares were surveyed each morning and each evening, when the crepuscular squirrels were most active. But the count went beyond just recording numbers. Squirrel behaviors and characteristics were noted, such as the squirrels' approximate age, vocalizations, coat coloration, and reactions to the volunteer squirrelologists. When the census was complete, the volunteers had recorded 2,373 eastern gray squirrels.

Even though no one is taking a survey today, the Squirrel Sighter app that sprang from that census can record Central Park squirrels or squirrels anywhere in the world. I open the app. One, two, both gray squirrels, and then I turn my attention from the squirrels to all the people around me who seem oblivious to the squirrels or are more interested in the pigeons and mourning doves pecking away on the ground in the shade of the trees, remembering that ubiquitous squirrels might be white noise to most, especially in this era of handheld screens and portable music.

At the turn of the twentieth century, when Central Park squirrels were having their moment, the park commissioner built them apartment houses to prevent them from being exposed to the harsh winter weather. The park's visitors and wealthy residents living nearby were already bringing them cracked nuts at that time.

Regrettably, the squirrels of today do not enjoy the remnants of those lavish gifts in the trees of Central Park's Ramble. Yet visitors can envision these gifts nestled in the wooded glens and the heated debates they ignited. New Yorkers, then and now, are known for their outspoken views, adding a layer of intrigue to the squirrel saga.

While many loved to pamper the park's squirrel population, others had strong words opposing the squirrel-housing gifts. "The effect of all this upon the squirrels has been weakening and demoralizing," wrote the author of *The Evening World's Daily Magazine* article "Squirrels and Men" in 1906. They were so beloved that they were even protected from cats and dogs. The author made the bold statement that these squirrels were not healthy. "They become fat through lack of exercise. Their fur is in bad shape through improper diet. Some of them have lung disease. All of them have lost their vigor and virility."

Looking at the healthy squirrels scurrying around the park, imagining distress over their welfare is difficult. The writer of that commentary, however, was correct to be concerned about human intervention. Feeding wildlife is problematic to them and their ecosystem.

While the squirrel apartments installed in 1907 are no longer a part of the park, many humans are now setting up squirrel homes and feeding perches on their properties worldwide. More than one popular photo of a squirrel nibbling a nut at a tree-mounted picnic table has crossed my social media feed. But do these feeders and homes fall under the auspices of intervening in the welfare of wild squirrels because we've caused them harm from habitat destruction? We are responsible for stepping up if we have caused habitat loss and an inhabitable environment. However, we must reflect on our behavior if we act solely for our amusement. Does wildlife exist exclusively to amuse us?

There are additional questions to consider, such as the current status of the Central Park squirrels since that census took place on our warming planet. Is anyone besides the park's visitors observing the squirrel population and recording data?

The 2018 Squirrel Census transformed into a captivating multimedia project, blending science, design, art, and storytelling to focus on the eastern gray squirrel. The findings were shared in various forms, including maps, 45 rpm records, and live presentations. The organizer's latest project, the Squirrel Census Phone Tree, continues this captivating storytelling by presenting stories and data from a March 2020 squirrel count of twenty-four New York City parks through a toll-free number.

We can thank the six-person staff behind this census project/squirrel entertainment, including the creator, writer, and humorist Jamie Allen, scientist Donal Bisanzio, and a canine named Sophie "Beanie Bear" Allen, who holds the title of founder. Sophie, an Aussie-Lab mix, was just two and a half when she happened upon a young squirrel in her backyard in Atlanta, Georgia. After smelling and tasting it, she was separated from the little bushy-tail. The sniffer dog hobbyist might have been capable of joining the Conservation Cattle Dog Squirrel Squad, working with conservationists in the Pacific Northwest, had she been trained; according to Allen lore, Sophie began her pursuit of gray squirrels with abandon. Her squirrel obsession spread to her human family.

Human Allen's latest project explored whether squirrels were plentiful enough to lead an uprising against reigning humans. Bisanzio had the science chops to coauthor "Evidence for West Nile Virus Spillover into the Squirrel Population in Atlanta, Georgia." Sophie had a fourteen-year career in sighting and chasing squirrels.

In October 2022, on the anniversary of the Squirrel Census and four years after Sophie's rainbow bridge crossing, Jamie Allen authored a post that appears on the website of the *Central Park Conservancy Magazine*, "Getting to Know Central Park's Squirrels." An editor's note preceding the post informs readers that the conservancy works to "prioritize native plant species that nourish our wild residents, remove invasive species, and leave snags (dead or dying trees) in the woodlands—all in order to cultivate a habitat in which the Eastern gray squirrel, and many other species, can find sustenance and shelter."

Methodology created by Danish American biologist Vagn F. Flyger inspired the 2018 census. After the two-week data collection period in the park, the findings were plugged into a Flyger formula to achieve the estimated population numbers—about 2.74 squirrels per acre.

Surprisingly, the study revealed that squirrels are not evenly distributed throughout the park but live in clusters, "like stars in a galaxy." The densest "constellations" were found in areas with thick tree canopies, such as the wooded Ramble, which had the highest squirrel density. The team recorded 155 squirrels, approximately 15.5 squirrels per acre, more than five times the park's average.

As in all urban parks, the Central Park squirrels exist on an island encircled by pavement and buildings. "It's a rectangular green haven in a sea of cement, steel, and killer cars," wrote Allen. The stark reality is, "if they cannot make it there, a squirrel version of Sinatra might sing, there isn't anywhere else."

Allen wrapped up his post by answering his most-asked question: Why did he create the Squirrel Census? "It's important to note that when you engage in any kind of census, you don't just learn about the animal you're counting; you also create a profile of—and better understand—the community or country in which it lives. Our census and all of the supplemental data we gathered, in other words, helped us perceive and appreciate Central Park—and all of its residents—in a fresh way." He concluded: "On a more poetic note, when you take a moment to get out in the world and look around, you just might see something that stays with you—maybe even something that reflects the personality of the city that we call home."

The Celestial Map that was created with the results of the 2018 Central Park Squirrel Census appears on the website, showing the clusters of squirrel sightings, fur coloration patterns, and other data. It was a popular item that sold out, along with the complete 2019 report.

Colin Jerolmack, professor of sociology and environmental studies at New York University, calls squirrels "pedestrian animals." According to him, they are pedestrian in both meanings of the word: "They're common, everyday animals . . . and they literally walk the sidewalks."

How could we comprehend the state of squirrels during this turbulent Anthropocene era of biodiversity loss without the invaluable contribution of community and scientific data? Community involvement is crucial as researchers struggle to fund their studies with the small pot of available federal money and the reality that fewer than 10 percent of their submitted grants are funded. Young researchers who might be eager to study a ubiquitous species might experience difficulty obtaining a grant if they haven't received one before. A proven track record is an important factor in grant acceptance. Place that consideration alongside the belief that squirrel research may not be as critical as other species research and it becomes even easier to understand why there are still vital gaps in our bank of knowledge about many squirrel species.

While gathering population data is important, whether from community members or scientists, it is also essential to understand how squirrels are surviving during this challenging period. It is vital for the health of all these species and the health of their ecosystems.

*Chapter Thirteen*

# As a Shape-Shifter

> "Darwin's theory of evolution is a framework by which we understand the diversity of life on Earth. But there is no equation sitting there in Darwin's 'Origin of Species' that you apply and say, 'What is this species going to look like in 100 years or 1,000 years?'"
> —Neil deGrasse Tyson

Squirrels, with a history on this spinning planet far longer than ours, are an intriguing subject to study. The oldest tree squirrel fossil, the *Protosciurus jeffersoni* [*Douglassciurus jeffersoni*, 1901]) skeleton, was a significant find. It can be seen, frozen in time, holding a fossilized walnut in its forepaws, in the Smithsonian Institution's Objects of Wonder collection. The critical skeleton, numbered PAL243981, is not much different from the collection's other mounted squirrel skeleton from 1880, a familiar gray that you could imagine scurrying across your local park. However, the *Douglassciurus*, a distinctive creature that lived in the present-day state of Wyoming in the Late Eocene Epoch 38–34 million years ago, had its own unique features. The Eocene, a period featuring the appearance and diversification of many modern groups of mammals and other organisms, is appropriately derived from the Greek word *eos*, "dawn." That early squirrel evolved from a roughly two-foot-long creature of the genus *Ischyromys*, a likely arboreal mammal that was a good climber.

The fossilized squirrel, a discovery that owes its existence to the keen eye of Jennifer Emry in 1975, is a testament to the human aspect of paleontological research. This significant find, from the White River Formation, which was once an aggregational meandering river system, now resides in the Smithsonian's museum case. When alive, it lived in a world of temperate and subtropical forests with trees and plants most similar to or identical to those found today in Central and South America. The early Eocene greenhouse climate was warm and humid.

The Smithsonian's squirrel specimen, identified by Robert Emry and Richard Thorington, lived long before our common eastern gray squirrels. Our grays appeared in the tree canopy of present-day North America a mere 4 million years ago. They and today's other squirrel species have been doing their squirrel thing ever since.

Scientists used dental research to support identification and the divergence between tree and flying squirrels about 23 million years ago. However, recent studies show that the two groups may share the dental features scientists have used to distinguish between flying and non-flying squirrels. This skeleton specimen reflects little morphological change in flying squirrels for almost 12 million years.

Paleontological and molecular studies of the skeletons found have demonstrated that flying squirrels evolved from tree squirrels as far back as 31–25 million years ago. The most extraordinary finding is that these results show that the extinct genus *Miopetaurista* is closely related to *Petaurista*, a modern group of giant squirrels that inhabit tropical and subtropical Asian forests. This modern group of giant squirrels could be considered living fossils.

Analyzing fossil data enables scientists to further their comprehension of living species and can give us clues to how they will respond to global climate changes today. Human-induced climate change caused by the burning of fossil fuels, deforestation, and farming of livestock has affected storm strength, wildfires, drought, and water security, providing stressors for humans and wildlife. Will these changes influence evolution in wildlife? Scientists are exploring the apparent shape-shifting of squirrel species during this changing Anthropocene epoch.

Two Duke University evolutionary biologists, John M. Mercer and V. Louise Roth, are among the scientists involved in this exploration. They analyzed the DNA differences among fifty of the fifty-one present-day squirrel genera to explore their interrelationships, using both live animals and preserved museum specimens. They left out only one elusive, rare Indian flying squirrel. They also explored fossil-dated records and "molecular clock" analyses, which predict how DNA sequences change over time. They found an interesting interplay between global change and how this particular group of animals diversified.

Frozen in time as if going about its day, the Smithsonian's squirrel skeleton captures a moment in evolution. The researchers determined that in the 5 million years that followed that fossil evidence, there "appears to have been a very rapid divergence of squirrels into five major branches." The timing of that divergence corresponds to an abrupt cooling, climate change, and many animal extinctions.

The pair unpacked their DNA study in a paper published in the journal *Science*: "By modifying habitats and creating bridges and barriers between landmasses, climate change and tectonic events are believed to have important consequences for diversification of terrestrial organisms." They continued, "The squirrel family (Sciuridae) is one of very few mammalian families endemic to Eurasia, Africa, and North and South America and is ideal for examining these issues."

Early evidence of squirrels in Europe dates to around 30 million years ago. While researchers don't know precisely how the squirrels arrived there, they know that they could have easily moved between Eurasia and North America at that time.

Now we are involved in another period of climate change—this time, we've caused it. Biologists are exploring how squirrels adapt to new and changing environmental conditions during this period of human-induced rapid environmental change (HIREC).

Renowned nineteenth-century naturalist Charles Darwin wrote about the evolution of flying squirrels in 1859 in *On the Origin of Species by Means of Natural Selection, or the Preservation of Favoured Races in the Struggle for Life*:

> Look at the family of squirrels; here we have the finest gradation from animals with their tails only slightly flattened, and from others, as Sir J. Richardson has remarked, with the posterior part of their bodies rather wide and with the skin on their flanks rather full, to the so-called flying squirrels; and flying squirrels have their limbs and even the base of the tail united by a broad expanse of skin, which serves as a parachute and allows them to glide through the air to an astonishing distance from tree to tree. We cannot doubt that each structure is of use to each kind of squirrel in its own country, by enabling it to escape birds or beasts of prey, or to collect food more quickly, or, as there is reason to believe, by lessening the danger from occasional falls. But it does not follow from this fact that the structure of each squirrel is the best that it is possible to

conceive under all natural conditions. Let the climate and vegetation change, let other competing rodents or new beasts of prey immigrate, or old ones become modified, and all analogy would lead us to believe that some at least of the squirrels would decrease in numbers or become exterminated, unless they also became modified and improved in structure in a corresponding manner. Therefore, I can see no difficulty, more especially under changing conditions of life, in the continued preservation of individuals with fuller and fuller flank-membranes, each modification being useful, each being propagated, until by the accumulated effects of this process of natural selection, a perfect so-called flying squirrel was produced.

Darwin's fascination with nature wasn't limited to finches, worms, and tortoises. Squirrels also held a place in his thoughts, despite the misleading impression from the illustration "Darwin and the Squirrels" found in Charles Frederick Holder's early biography of Darwin, *Charles Darwin: His Life and Work*, which depicts squirrels climbing up Darwin's legs and watching him from a tree. The illustration was meant to portray an individual so immersed in nature that he was oblivious to the creatures around him. However, Darwin's profound connection with nature, as evidenced by his detailed writings about squirrels, was far from oblivious.

Darwin's observations of squirrels in *Origin* provided an example of how squirrels were adapting and changing in response to a changing climate rather than perishing. "Let the climate and vegetation change," he wrote, leading to his conclusion that under "changing conditions . . . each modification being useful, each being propagated" led to the "perfect so-called flying squirrel." His observations are relevant today as we witness physical and behavioral changes in wildlife during another period of changing climate conditions. Darwin wrote that "some at least of the squirrels would decrease in numbers or become exterminated, unless they also became modified and improved in structure in a corresponding manner."

A humorous November 2024 article set on the Maryville College campus in Tennessee focused how campus squirrels were preparing for unseasonable heat. "'We're sweating our tails off!' Nutwrencher remarked," wrote author Hannah Lee. "'We should be halfway through hibernation prep, but with these temperatures, we're still in summer mode. The whole thing's nutty!'" As humorous at this piece is, it addresses a real concern about squirrel survival amid increasingly unseasonable high temperatures. Lee reported that students on campus had "witnessed hordes of squirrels shoving acorns and walnuts into every nook, cranny, and even abandoned textbooks."

Fluctuating environmental conditions and increased human population may also be behind the recent discovery of California ground squirrels hunting and eating voles. While we've known that squirrels are omnivores and often are predators of nesting birds, this behavior had not been witnessed before. According to the research, "direct study of hunting behavior by squirrels remains rare," making this widespread carnivorous behavior, a dietary adaptation, groundbreaking.

Was Darwin correct? In addition to changing their behavior, are squirrel species physically modifying to survive this fluctuating period? Researchers claim yes.

While some species of squirrels might emigrate, as Darwin wrote, to places where living might be easier, one recent investigation in South Africa found a squirrel species that may be adapting physically to Earth's increasing temperatures rather than emigrating or becoming exterminated. Put aside the Unbeatable Squirrel Girl superhero persona; actual squirrels might be developing their superpower to survive in our warming world.

Researchers from multiple universities and countries have spent countless hours in the African heat studying the squirrel species. The S.A. Lombard Nature Reserve in South Africa is one of three nature reserves in and around the Bloemhof Dam where the Cape ground squirrel lives. Along with squirrel studies, this reserve is the site of pioneering conservation work on the black wildebeest. Fundamentally, Kalahari grassland is on an uncharacteristic floodplain; the reserve is virtually treeless, so there is not much shade.

The Cape ground squirrel is one of four African ground squirrels (genus *Xerus*) that inhabit savannas and rocky deserts in northern, eastern, and southern Africa. They spend about 70 percent of their lives foraging for food, which provides them with most of the moisture they need. They consume a minimal amount of water outside of that. How do they keep cool, then, while they forage for food?

Two of the researchers, conservation ecologist Miyako Warrington and her colleague Jane Waterman, observed Cape ground squirrels (*Xerus inauris*) in one of the nature reserves where the maximum daily temperature had increased by 2.5°C (36.5°F) in the previous two decades. They uncovered that the bodies of Cape squirrels had become smaller and their feet larger during this period of increasing heat. In 2022, the pair wrote in the *Journal of Mammalogy* that there is growing evidence that climate change is causing the "ecology, life histories, and physiology" of animals, including squirrels, to change.

Squirrels are not the only species on the planet that has decreased in body size. Studies have found that many bird species have reduced in size over the past three decades. During the nineteenth century, German biologist Carl Bergmann proposed the principle that there is a correlation between external temperature and body surface in warm-blooded animals. His principle became known as Bergmann's rule and is demonstrated in the difference between bulkier birds and mammals in cold regions and those with slighter sizes in warmer regions. Scientists have discovered that animals are shrinking in various places around the world in response to the changing climate. In a 1976–99 study, mountain wagtails, slender birds living across sub-Saharan Africa, were found to be lighter along the Palmiet River in Westville, KwaZulu-Natal, where the temperature had increased by 0.18°C (32.32°F). The lighter birds had replaced the heavier individuals in their population over time.

One study found a 0.6 percent body mass decline across 105 North American bird species, including a 2.8 percent decrease in the size of tree swallows and a 1.2 percent decrease in the size of our familiar harbingers of spring, American robins.

Squirrels already have many tools in their toolbox for dealing with heat, although sweating is not one of them. They use their bushy tails for a bit of shade. Cape ground squirrels do raise their tails to cool their skin. They also retreat into their burrows to cool off, another traditional way for ground squirrels to escape heat. Other squirrels, such as flying squirrels during a hotter-than-usual summer in Chattanooga,

Tennessee, seek out cool places in the nooks and crannies of area homes.

Squirrels also practice the behavior of heat dumping, or "splooting," the pose of sprawling flat on the cool ground that often puts a smile on our faces. Splooting is one way animals can shed heat through their less furry bellies, which contain more blood vessels. However, rising temperatures make this behavior more difficult. Just as humans find it more difficult to have sweat cool their bodies during high humidity, higher temperatures make it more difficult for squirrels to find relief by splooting. Without other ways of lowering their body temperature on this warming planet, squirrels will have to adapt, change, or perish, just as Darwin wrote.

The researchers in Africa lured the ground squirrels to traps baited with birdseed and peanut butter. After the squirrels were captured, they were weighed, measured, and released. Warrington and Waterman found that the squirrels' bodies had changed over an eighteen-year span. The squirrels had an increase in their already large feet, which had grown by about 11 percent in relation to their body size. In addition, the squirrels' spine lengths had shortened by about 6 percent.

"Cape ground squirrels are ecosystem engineers," noted Waterman. "They change the vegetation around their burrow clusters and increase grassland biodiversity, so changes in their body size and behavior can have consequences to the biodiversity of temperate grasslands, which are one of the most endangered ecosystems on the planet."

However, the researchers made it clear that their studies still left questions. They wrote, "We saw a pattern of larger feet and shorter spines at the same time that temperatures increased at our field site. This means that even though we saw these body changes, we cannot yet tell if these changes translate into the squirrels being better able to keep themselves cool."

The changing bodies of squirrels ignite new interest in studying other mammals during this global change. Investigating the effects of climate change on squirrels and other species is paramount.

"As species that are able to respond quickly to environmental changes are more likely to survive the rapid human-driven land-use changes and climatic changes projected for the coming centuries, understanding how species and populations respond to change in climatic conditions is important in predicting which species will be resilient to climate change," wrote Warrington and Waterman.

Researchers have already seen how squirrels can adapt and migrate. In a 2009 study, Canadian researchers explored the expansion of southern flying squirrels into their northern range limit, providing the opportunity for the formation of hybrid zones between two species of flying squirrel, *Glaucomys sabrinus* and *G. volans*. This was the first report of hybrid zone formation induced by climate change.

Another study by Canadian researchers focused on how climate change alters squirrels' sperm. University of Manitoba researchers studied male Richardson's ground squirrels (*Urocitellus richardsonii*) found in Canadian prairie habitats. They found that when the squirrels emerged from hibernation earlier than usual during an unseasonably warm winter, their sperm was nonmotile, or incapable of movement. In other words, the squirrels were "shooting blanks."

Arctic ground squirrels (*Spermophilus parryii*, also referred to as *Urocitellus parryii*) have adapted to the harsh weather of Arctic winter months by hibernating for about eight months. However, climate change has altered that behavior, and they are emerging from hibernation earlier these days. A twenty-five-year research project found that hibernation periods are growing shorter, and there are differences between male and female hibernation periods. The report shows a direct link between temperature changes and the ecology of squirrels. Researchers found that female squirrels emerged from their burrows before their male counterparts. This behavior change could affect their ecosystem food web both positively and negatively. Since the females do not need to use as much stored fat during hibernation and can begin foraging for food earlier, this behavior could lead to healthier litters and higher survival rates. That's the good news. However, if males don't shift their behavior too, the females might be at greater risk of being consumed by a predator, like a hungry grizzly, leaving the males with less mates. This shift could create an imbalance in the overall squirrel population. Hence the bad news.

The impact that a potential decrease in squirrel numbers has on their ecosystem can also be seen in other studies. John Koprowski and Mike Stokes have studied squirrels in South Africa to observe how they, and other rodents, are regenerating forests where elephants are causing damage. In China, Koprowski has studied squirrels and other seed-eating rodents as an indicator of climate change at high elevations where giant pandas live. "Overall, we are interested in using squirrels as indicators of forest change and climate change over time," said Koprowski.

University of Minnesota doctoral researcher Amy-Charlotte Devitz is another scientist who is focusing on how squirrels adapt to this challenging period. She has been working with Emilie Snell-Rood on how organisms face environmental change on a scale unknown in their evolutionary history. Unlike many other researchers, Devitz has been surveying squirrels in urban environments. She and her team collected squirrels from nine different sites around the Ann Arbor area of Michigan in 2020, ranging from remote areas to the heart of the University of Michigan's campus. "In the urban environment there are, relatively speaking, evolutionarily new stressors. Humans, cars, dogs, artificial light, trash, food waste, these are all distinct to the built environment, and we want to see how, and to what extent, they may be altering the behavior and personalities of the animals we share it with."

Probing a familiar, "pedestrian" species, like squirrels, during this critical global period is essential to grasping how every other organism reacts to the changing environment. Consider that squirrels can act as an indicator species for ecosystems, replacing the canary in the coal mine. Continuing studies on squirrel evolution should spark further vital studies on other species, their ecosystems, and how humans are responding. As with any species undergoing stressors, Darwin's thoughts on whether the species changes, moves, or disappears are even more relevant today. While we discover that some squirrel species are changing, still others are disappearing.

*Chapter Fourteen*

# As an Endangered Species

PALM SQUIRREL.

> *"There are some four million kinds of plants and animals in the world—four million different solutions to the problems of staying alive."*
> —David Attenborough

While naturalist Ernest Thompson Seton wrote and spoke about the importance of squirrels to our character and ecosystems, he also voiced strong opinions on their fate: "The greatest enemy of the Gray-squirrel is undoubtedly the white-man, armed—not with the rifle, but with the axe. While the great basin of the Mississippi Valley was crowded with nut trees that never failed of a bounteous annual supply, the Squirrel could scoff at the riflemen, be they never such dead-shots. But when the forest itself was laid low, and the possibility of new generations cut off, the Squirrel hordes were like a river whose parent springs were dried up at the source."

His words have never been as accurate as they are now during this human-powered Anthropocene epoch. Not all squirrel species are evolving like the Cape ground squirrels, moving or being translocated to safer or greener pastures, like prairie dogs, to survive this challenging world landscape. Unfortunately, many squirrel species are instead experiencing plummeting population numbers in ever-decreasing habitats falling to the axe and human-induced rapid environmental change.

In fact, according to Baldwin Wallace University biology professor and squirrel expert Karen Munroe, about 80 percent of the world's squirrel populations are threatened or endangered during this period of global warming. Wildfires, floods, droughts, and other changes to squirrel habitats have affected squirrel populations. In addition, our view and management of squirrels and their habitat has influenced many species. That statistic of 80 percent should surprise us, but as she reports, "It's really only here, in the US, that we think of squirrels as pests, and a species to be managed and hunted." That view has made an impact.

Munroe has been investigating squirrels since she was seventeen years old. She spent her graduate research exploring round-tailed ground squirrels' social behavior and mating systems. She brought her observational skills, genetic and molecular techniques, and curiosity to her university students in Ohio, inspiring the next generation of squirrelologists as they work on their own projects.

She speaks about a squirrel culture that focuses more attention on our common native squirrel species and often diverts our attention from the additional species around the world and the countries where they face increasing challenges to their survival, from habitat loss to climate change to invasive species. According to the university's Squirrel Team leader, many squirrel species, such as one in Japan, are living in isolated populations: "The only places you find squirrels anymore in Japan are the really old, sacred shrines because those are the only places left that have enough large, old-growth trees to support a population of squirrels."

We began seeing how habitat fragmentation threatened a North American squirrel population in the early days of the Endangered Species Act of 1973 (ESA). In 1967, the Delmarva Peninsula fox squirrel (*Sciurus niger cinereus*) was officially listed in the act's precursor, the Endangered Species Preservation Act. The silver-gray-coated squirrel with short, rounded ears and whitish feet resembles the more common, smaller gray squirrel. Faced with habitat loss from development and forest clearing, the population was in a deep dive. Fortunately, the listing required a recovery plan. This one worked. Successful efforts to rebuild the endangered squirrel population enabled removal of the species from the endangered list in 2015. Although not common, the squirrel is now considered a recovering endemic subspecies. Its range is in the Eastern Shore of Maryland and a few spots in Virginia and Delaware. One small population resides in Maryland at Third Haven Woods, a Nature Conservancy property, a remnant of the forests that once covered much of the Delmarva Peninsula.

Even though it is surviving in only a small percentage of its historical range, the Delmarva Peninsula fox squirrel was saved from extinction. Other squirrel species have not been so lucky. Still others are struggling.

The Carolina northern flying squirrel (*Glaucomys sabrinus coloratus*), listed under the ESA in 1985, inhabits a limited region in western North Carolina, southwestern Virginia, and eastern Tennessee. The high mountaintops in the region's Cherokee lands are a crucial stronghold for its recovery. As with the Delmarva, habitat loss is a threat to this species. Logging and the loss of balsam trees to a tiny, invasive sucking insect, the balsam woolly adelgid (*Adelges piceae*), have taken a toll, but perhaps the biggest threat to this species is climate change. According to the US Department of Agriculture, the "shortened season of warmer temperatures that characterize some of the higher elevations, especially in the northern latitudes, can either result in fewer generations per year, or possibly limit the distribution of the adelgid entirely."

North Carolina's warmer temperatures, however, enable a third generation. A partial fourth generation occurs in the lowland areas of Oregon and Washington. Will projected warmer temperatures affect fir forests and squirrel populations that are not currently affected? A 2022 estimate found about 33 percent of the Nantahala and Pisgah National Forests in North Carolina that once had spruce and fir trees in the canopy were dominated by other species. This finding strengthens the need for spruce-fir forest restoration, but the changing climate makes this effort challenging.

According to the US Fish and Wildlife Service's 2022 5-Year Review of the Carolina Northern Flying Squirrel Recovery Plan, the recovery criteria needs to reflect the best available and most current information on the biology and habitat of this species. Using traditional labor-intensive, invasive live capture methods isn't sufficient.

Those traditional methods are well-known to wildlife biologist Craig Stihler, who had studied the tiny northern flying squirrels since 1985, when this subspecies was listed as endangered under the ESA. He's probably held more West Virginia northern flying squirrels than anyone. Not much was known about the species when he began his studies, other than the basics—what the squirrel ate, where it slept, and how it differed from its common cousin, the northern flying squirrel. The fact that there were so few of these feisty, mysterious, nocturnal animals made it even more challenging to get at the heart of why it had declined and how to save it. Delisting the subspecies in 2013 did not make it common. Seeing one in the wild is still a rare sight. More methods of study were needed in order to recover the population.

However, investigating mysterious nocturnal flying squirrel species is always challenging. Fortunately, for her dissertation at Auburn University, Michelle Gilley discovered another way to study them. She uncovered a way to characterize the ultrasonic calls of North American flying squirrels. Her new method presented a way forward for flying squirrel research, but it wasn't perfect, nor would it solve the challenges for the species.

Despite the significant strides made in exploring flying squirrels, particularly the influential work of another wildlife researcher, Corinne Diggins, and her team, who utilize ultrasonic acoustics to detect these cryptic and elusive squirrels for the Department of Fish and Wildlife Conservation at Virginia Tech, the need for further research remains pressing. Effective monitoring methods are still a necessity, underscoring the ongoing importance of this methodology.

Although climate change remains one of the biggest threats to the Carolina northern flying squirrel, it was not addressed in the squirrel's recovery plan or criteria. In addition, the effects of different timber management techniques on this species have yet to be determined. This is hugely problematic and indicates another area where more information is required for the recovery of this "cryptic" animal.

The Carolina northern flying squirrel's companion on the ESA list, the Mount Graham red squirrel (*Tamiasciurus hudsonicus grahamensis*), appeared for the first time in the December 30, 1982, *Federal Register* under the US Fish and Wildlife Service's report of endangered and threatened wildlife, stated December 17, 1982. The tiny squirrel subspecies with an eight-inch body and a six-inch fluffy tail was listed between Sherman's fox squirrel (*Sciurus niger Sherman*) and the northern flying squirrel (*Glaucomys sabrinus*). It was officially listed as endangered five years later, on June 3, 1987, in a report authored by Alisa M. Shull, an endangered species biologist with the US Fish and Wildlife Service. "The Service determines endangered status for the Mount Graham red squirrel, *Tamiasciurus hudsonicus grahamensis*, a small mammal found entirely in the Pinaleño Mountains of southeastern Arizona," she wrote. "Its isolated habitat has declined over the last century and may face additional losses to logging, recreational development, and construction of an astrophysical observatory. The red squirrel may also be in jeopardy because of its reduced numbers and through competition with an introduced species of squirrel."

Isolated from other red squirrel subspecies since the end of the Pleistocene glacial periods, Mount Graham red squirrels were thought to have been extinct in the 1950s. That changed when small numbers were discovered in the 1970s.

John Koprowski, professor emeritus at the University of Arizona, has studied fire and insect damage in high-elevation coniferous forests that influence the population of the Mount Graham red squirrel (*Tamiasciurus hudsonicus grahamensis*) since 2000 as part of the Mt. Graham Red Squirrel Research Program, which he conducted out of the Conservation Research Laboratory of the University of Arizona until 2020. He then moved to the University of Wyoming as Wyoming Excellence Chair and dean of the Haub School of Environment and Natural Resources.

During the past two decades of his study, he has seen additional forest losses that continue to fragment the squirrel's habitat, including substantial wildfires that burned 12,029 hectares (about 30 acres) in 2004 and 19,600 hectares (about 47 acres) in 2017. The 2017 Frye Fire destroyed almost 50,000 acres on Mount Graham. It was not clear how many endangered reds survived, but the number estimated was dangerously low, at thirty-five individuals. This led the Center for Biological Diversity to file a lawsuit in March 2024 in the US District Court in Tucson, alleging that the US Fish and Wildlife Service had violated the ESA by failing to update and expand the endangered red squirrel's habitat after a 2017 petition.

"It's pathetic that the [US] Fish and Wildlife Service recognized these squirrels were in jeopardy more than two decades ago but now ignores this when there are fewer squirrels and less habitat," said Robin Silver, cofounder of the Center for Biological Diversity. "The squirrels' situation is now even more dire as they face a historic habitat bottleneck, isolated in tiny islands of mature canopied forest that are widely separated from each other. This greatly increases predation risks and the likelihood that these squirrels will go extinct."

However, the fires have also been a threat to the endangered squirrel. The construction of the Large Binocular Telescope, located on

10,700-foot Mount Graham in Arizona's Pinaleño Mountains, is also a threat. The construction has continued since 2004 at the Mount Graham International Observatory and has disturbed the endangered squirrel's habitat with electric power lines. In addition, insect outbreaks contributed to significant tree mortality of spruce-fir forests in the monitored habitat, providing yet another threat. A perfect storm for population demise existed.

But at the end of 2024, after concerted conservation measures that included habitat assessment, conifer seed collection, insect pheromone treatments to protect trees, forest stand monitoring and enhancement, a captive rearing program, and annual survey monitoring, there was good news for the endangered subspecies. Although the Frye Fire had caused the Mount Graham population to drop drastically, the US Fish and Wildlife Service reported numbers climbing from that dangerously low number of 35 squirrels after the fire to 144 individuals in 2023 and then to an estimated 233 reported in December 2024. Staff of the Arizona Game and Fish Department were surprised by the increase they found in squirrel and midden numbers. The rebound is fantastic news and highlights the resilience squirrels can exhibit when protections are in place.

"This survey process allows us to obtain a more accurate picture of the population numbers," said Kerwin S. Dewberry, forest supervisor at the Coronado National Forest. "We are excited to see that current land management practices being implemented in the Pinaleño mountain range are proving to be beneficial for the Mount Graham red squirrel population."

Recovering endangered and threatened squirrels requires collaboration. The Phoenix Zoo keeps a handful of endangered Mount Graham red squirrels to study and use for a pilot breeding program that will enable young to be released in the wild. The zoo is the sole institution housing a small captive population and is partnering with the US Fish and Wildlife Service, the University of Arizona, the Arizona Game and Fish Department, and the USDA Forest Service to conserve the species.

The Phoenix Zoo's director of conservation and science, Tara Harris, feels that Mount Graham red squirrels are "one of the most interesting species" that she works with out of all of the ones in her purview. On my visit to her facility, she told me that it isn't that she's more passionate about squirrels than the other species she works with, but she loves a good puzzle, a good challenge. "They are one of the most challenging species I've worked with because we're writing the book on how to do this [recovery] and we haven't been successful yet."

The pilot breeding program began over ten years ago. At home on the mountain, the squirrels live in a cold environment. Despite replicating the conditions and food as closely as possible and conducting extensive observation, the program has yet to produce any young squirrels, but Harris isn't giving up. "I want to figure it out. I want to be successful and be able to return them to the mountain."

The squirrels' survival is at stake. "We have such lofty goals, but still they elude us in terms of what it is going to take to be successful. I want us to be able to use our brains as scientists and just figure it out."

The sounds of the squirrels scrambling in their enclosures near her office is a constant reminder of the puzzle that needs to be cracked for their survival. While Harris is a formidable force of perseverance, endangered species recovery has its share of disappointments.

Along with the Carolina northern flying squirrel and the Mount Graham red squirrel, already officially listed under the ESA, the western gray squirrel (*Sciurus griseus*) was recently formally listed as endangered by the Fish and Wildlife Commission in Washington state after wildfires in the region left it with greatly diminished habitat. "Sadly, western gray squirrels are barely hanging on to existence," said Pat Arnold of Friends of the White Salmon River in November 2023.

There are three remaining populations of the western gray, the largest of the state's gray squirrel subspecies. It has a fluffy, often called beautiful, body-length tail. The entire population, isolated and critically small, numbers barely around four hundred. Like other threatened species, it faces a number of threats from anthropogenic climate change and habitat loss from logging and development.

In early 2024, the Washington Department of Fish and Wildlife put out a call for information from the public, including nongovernmental organizations, universities, private researchers, and naturalists, to help inform the state of the status of two species of threatened ground squirrels in the state, the Washington ground squirrel and Townsend's ground squirrel. It is hoped that this information will provide data on the species' demographics, current habitat conditions, threats and trends to populations, and existing conservation measures that have benefited the species.

It was too little, too late for Friends of the White Salmon River, which joined together with Vancouver Audubon, Friends of the Columbia Gorge, the Center for Biological Diversity, and WildEarth Guardians to file suit in Washington's Clark County Superior Court on February 23, 2024, against the Washington State Department of Natural Resources and public lands commissioner Hilary Franz for failing to protect habitat for the endangered western gray squirrel.

The lawsuit claimed that the western gray is an indicator species for the "health of rare oak woodland habitat" in the state and has been allowed to decline for decades with the lack of protective regulations. What is the future for this endangered squirrel species? How should we consider this population or other squirrel populations as we further fragment their habitat and witness climate change implications?

Similar to Central Park's squirrel population, these wild squirrel species, although not living in urban settings, also live on habitable "islands." Misunderstanding these island populations is comparable to climate change deniers not believing that the climate is warming because they experience cold in their location. Squirrel population numbers might be the same. The squirrels you see in your backyard, village, or city park don't represent an entire squirrel population. The same can be valid for a lack of squirrels. Without scientific studies, we don't know the realities of these populations.

The western gray squirrel (*Sciurus griseus*) was deemed threatened in 1993. Large-scale wildfires in 2014 and 2015 wiped out large populations, destroying tens of thousands of acres of squirrel habitat. As wildfires continue to increase, squirrels of various species are increasingly threatened. They are also susceptible to diseases such as mange and mosquito-spread western equine encephalitis, which could spread more quickly with rising temperatures. Although most cases have occurred west of the Mississippi River, in the late summer of 2024 cases of western equine encephalitis were reported in New England and resulted in one human death. As of this writing, no squirrels have been reported with the illness.

Placing a threatened squirrel species, such as the western gray squirrel, on a state's endangered list is one of the first attempts to restore its population. A 2007 plan requested that landowners voluntarily preserve ponderosa pine and Douglas fir trees, but that wasn't successful in helping the species. Even with this action, habitat loss continues. What about other squirrel species?

Another western squirrel species, the Washington ground squirrel (*Urocitellus washingtoni*), is currently threatened. In this case, officials have turned to detection dogs to obtain data about their population. For about ten years, a pack of squirrel-detection dogs from Rogue Detection Teams has sniffed out threatened Washington ground squirrels that inhabit the shrub-steppe prairie ecosystem on the border of Oregon and Washington. Teamed up for a project with The Nature Conservancy, this Conservation Cattle Dog Squirrel Squad, including Filson, Dio, Whisper, and Jack, conducts probably the most challenging task for any dog. Rather than give in to their canine instinct, the well-trained detection dogs have refrained from chasing these scattering ground squirrels; rather, they have noninvasively sniffed over forty thousand acres for signs of squirrel burrow activity, providing essential data to land managers and biologists. As a team, the dogs and their human partners, called bounders, cover miles of terrain each day, while avoiding snakes and prickly plants, to contribute to the future of the squirrel species.

Bounder Collette Yee grew up in the metropolitan San Francisco Bay Area. Today, Yee and her cattle dog, Jack, are members of the Rogue unit. Jack, a rescue dog, didn't have a great start in the field. In fact, like some seemingly unmotivated children, he was labeled a problem. It took a little extra work, but with Yee, Jack has found his groove. He needed to get out in a new place and have a job. As I write this, it's spring, and he's up early with Yee and out in the field using his nose to detect the activity of these endangered squirrels.

The window for ground squirrel detection work is small. The team heads into the varied landscape of juniper, sagebrush, and prairie during the spring, when the squirrels are active, and before the summer heat is problematic for the working dogs and potential wildfires threaten the teams. By autumn, the squirrels are again underground.

Yee has seen a good increase in the endangered squirrel's population since she began her fieldwork. "My first year, in 2019, out there we were finding burrows, but it didn't seem like a lot." The previous winter had been harsh and could have affected the population. In 2023, things were much different. "So this is what it could look like out here. It was so cool to see how active they were."

Yee enthusiastically described the burrow network she sees as looking like "a town with well-worn trails and roads between the burrows. And you can see the main hubs. There will be a big sage with several burrows entering into it. Roads going all out into smaller little burrows. It's so fun to see."

The ground squirrels are not as easy to see as their trails and burrows. "They are really good at keeping themselves hidden," said Yee. The second the bounders walk into the area with the dogs, the squirrels begin vocalizing to warn one another of the intruders.

Yee and other bounders have much to say about why we should care about these endangered ground squirrels. "There are species that the more you get to know them, the deeper the appreciation of how interesting they are. Every time I am out there I come out with more research questions."

Beyond their squirrel colonies, the Washington ground squirrels are part of a much larger community across the grasslands. The dug burrows Yee monitors and those beyond are used by snakes, kangaroo rats, and other animals. Like other squirrels vital to their ecosystems, Washington ground squirrels play an important role as they actively move nutrients and seeds across the landscape, and they are an important food source for raptors, such as the ferruginous hawk, and badgers.

Is more research possible?

"I would love to do more research on them," said Yee. "I really think that if more people saw how incredible their colonies are and the way they live together, it would get people much more interested. It feels so relatable. It's like watching a little town of people, except they're cute."

She added, "It really blows my mind when I think about how every small squirrel is a part of this network that makes a colony, and how interconnected the entire grasslands are."

In the meantime, the data that Yee is collecting is used in attempts to discern the occupancy and distribution of Washington ground squirrels in this corner of the Pacific Northwest. During August's fire season, I spoke with Kelly Wallis, who works on the other side of Yee's data collection. After a short update on the local season, which involved big and little fires at the area preserves, we moved the discussion to ground squirrels and her work.

Wallis grew up across the country, in a place devoid of squirrels in Georgia. But just a short ride away, a tree squirrel population flourished at the home of her grandparents. During her childhood, she and her family enjoyed the squirrel antics so much that one year they trapped some and brought them to release at their own home. Soon Wallis's family and everyone else nearby could be entertained and agitated by their own squirrel population. We both had a laugh as Wallis shared those early squirrel memories with me and how she had experienced those joys and also the frustrations of having her car wires chewed. And again, I was reminded that everyone has their own squirrel story or stories to share.

"People do think that they are almost a target practice species and a nuisance species. There are a lot of misconceptions about squirrels in general. And there is not a lot of appreciation. They are either these cute cuddly park things or something to shoot at," said Wallis in a distinctive relaxed Georgia cadence.

Wallis, who loves being in the field, now works with The Nature Conservancy in north-central Oregon with the data Yee gathers on Washington ground squirrels in the Boardman Conservation Area property that the conservancy manages.

Wallis recounted that the Washington ground squirrel population study began decades ago, in 1998, before the sniffer dogs' involvement, with expensive pedestrian data collectors. When the program evolved to incorporate detection dog teams, it was found to be more economical and thorough. Wallis corrected me when I called the data collecting a study. "You are right. Squirrels are very understudied," she said.

"And even with our ongoing monitoring, I wouldn't call what we're doing a study of them; it's just more that we're keeping track of what their colony size and their distribution is—just in our one little neck of twenty-three thousand acres, which sounds like a lot, but in the grand scheme of things it is a very small area."

Our discussion was interrupted by a violent storm on my end, and we reconnected after a little while as thunder still boomed outside. The storm, however, moved our focus to how these threatened ground squirrels were faring during this period of fires and wild weather. Certainly, habitat loss from agriculture replacing grasslands adversely affects the ground squirrel population as squirrels switch from natural grassland food sources and are challenged by having fewer areas to create their burrows. In 2019, an abnormal winter event took place that led to snow in March, much later than usual. The squirrels had to adapt by feeding on a different grass species in order to survive. It is unknown how these abnormal weather events will challenge the threatened species in the future.

Wallis was later able to supply an update on the surveys we discussed, which showed that the detections for Washington ground squirrels rebounded from the low numbers in the 2019 surveys to the levels surveyed before the 2019 winter. "Squirrels are resilient!" she added.

As we wrapped up our discussion, Wallis reminded me that the data gathered doesn't give definitive population information. There aren't any DNA studies occurring that would supply data on the individual squirrels. This is yet another example of how the lack of funding affects our knowledge base about a species—threatened or common.

The new recovery plan protected the Washington ground squirrel. It slightly increased penalties for killing endangered squirrels and initiated new regulations for the Forest Practices Board to consider forest practices that affect the squirrel's essential habitat.

Another North American species requiring data is the Arizona gray squirrel (*Sciurus arizonensis*), which was last assessed on September 20, 2016. It is native to Sonora, Mexico, and the US states of New Mexico and Arizona. It faces threats from hunting, fire suppression, and invasive species. North America has numerous squirrel species of concern, including several species in Idaho.

Yet another, the Mohave ground squirrel (*Xerospermophilus mohavensis*), was listed as a threatened species by nearby California's Fish and Game Commission as early as 1971. It is just another example of a squirrel species that can become vulnerable. With urban expansion, off-road recreation, agricultural irrigation, and military base expansion, the Mohave ground squirrel no longer exists in over 40 percent of its historical range. Climate change has added new challenges, with more extended periods of drought and higher temperatures. Petitions for the ground squirrel to attain federal protection under the ESA have gone on for decades, the most recent being submitted by Defenders of Wildlife and other conservation organizations in December 2023. This ground squirrel is the smallest in North America and inhabits solely the western Mojave Desert.

Like all squirrel species, the Mohave ground squirrel occupies an essential niche in its ecosystem. It aerates soil as it digs burrows, and it helps disperse native plant seeds, including seeds of the iconic Joshua tree, also threatened by climate change and wildfires.

Is it possible to reach a tipping point now that squirrel habitat is changing and decreasing worldwide? Although squirrels have their devoted fans, few scientists and government agencies are surveying their populations, compared with efforts on behalf of other species. However, Alice Morris, a researcher from the University of Idaho, is studying the decline of the northern Idaho ground squirrel, which has lost more than 99 percent of its historical range. The leading hypothesis for the decline is that a lack of fire has enabled encroachment and degraded the squirrel's habitat. Morris believes that if the researchers discover that the squirrels need forested areas to hibernate, it might lead them to rethink how the state manages its forests. Considering our squirrel populations can encourage other ways to keep our environment healthy for all species. Fortunately, the northern Idaho ground squirrel project has received funding from the USDA Forest Service, the US Fish and Wildlife Service, and the Idaho Department of Fish and Game because the species was listed as threatened under the ESA in April 2000. However, many states do not have the appetite or economics to produce the studies needed to protect our biodiversity for species that are not threatened or endangered. Most state agencies, like

New York's, rely solely on hunters for population and distribution data of squirrel species. In addition, the US Fish and Wildlife Service has minimal information about most squirrel species, except for the four species officially listed. This indicates the need for university researchers to initiate studies that will encourage a greater comprehension of our global and mysterious squirrel species.

However, there is an entire world of squirrels beyond the endangered and threatened species in North America that are facing similar threats. Of the squirrel species listed by the IUCN, at the time of this writing 93 demonstrate a stable population trend, 72 are decreasing, 4 are listed as increasing, and 116 species' population numbers are unknown. While most squirrel species are listed as Least Concern, the data also shows that scientists do not know the population trends for 40 percent of global squirrel species during this period of human-induced rapid environmental change and increasing habitat loss. About one in every five squirrel species shows an elevated conservation risk or needs more information for its assessment.

The IUCN lists 2 squirrel species as Critically Endangered, facing an extremely high risk of extinction in the wild; 15 species as Endangered, facing a very high risk of extinction; 16 species as Vulnerable, facing a high risk of extinction in the wild; and 23 species as Near Threatened, not currently threatened but could become so. There are 194 species listed as Least Concern. In addition, 35 species are data deficient.

The list of species lacking data outside of North America includes two with known decreasing populations: Bangs's mountain squirrel (*Syntheosciurus brochus*) and Hagen's flying squirrel (*Petinomys hageni*).

A baby palm-size Asia Minor ground squirrel was found in Turkey's Erzurum, seeking safety from the scorching summer heat plaguing the region. The Asia Minor squirrels, also called Anatolian ground squirrels, are found throughout Turkey, mainly in central and eastern Anatolia. Their population extends into the countries of Iran and Armenia. This little one made the news in 2023 and is representative of many others searching for refuge in abnormally warm temperatures worldwide. But human-induced climate change isn't the only factor threatening squirrel species.

As urban centers grow globally, pollution also becomes increasingly challenging, especially for wildlife such as squirrels. While the 2018 Squirrel Census found that the Central Park squirrel population was healthy in its numbers, another study found that the physical health of urban squirrels in Great Britain is threatened by urban air pollution. The deteriorating air quality in London has not uniquely affected humans but has also influenced the health of the city's gray squirrels. While pollution was not part of the plan to rid the United Kingdom of its unwanted gray squirrel population, the finding was made possible by culling the nonnative grays.

Although there were no existing pollution studies of squirrels, there were studies of free-living populations of feral dogs, pigeons, and a Brazilian rodent known as the tuco-tuco. Gray squirrels are an ideal model for analysis because they inhabit all of London's green space, which means they are exposed to realistic and complex ambient air pollution levels. This exposure is "something that human correlative studies and lab-based animal experiments rarely achieve." The research, published in the journal *Environmental Pollution*, highlighted airborne contamination in cities and suggested that squirrels, along with other wild species and domesticated animals, could be affected by pollution particles. "The current gap in our understanding of how wild populations are affected by and respond to TRAP [traffic-related air pollution] toxicity hinders our ability to effectively monitor, manage and predict an emergent risk to the health of all organisms."

The team of six researchers studied 106 squirrels from central London boroughs—Camden, Greenwich, Haringey, Richmond upon Thames, and Westminster—as well as two rural sites. The squirrels were studied for symptoms of lung disease, the presence of black carbon particles in their lung tissue, and damage to lymph node tissue.

Researchers found that squirrels from the inner city had more black carbon in their lung tissue but less lymphoid tissue than those culled from London's outer areas. The general finding indicated that the lungs of the squirrels in central London, where there are fewer trees, had suffered from more significant air pollution. Squirrels living in leafier parts of London lived longer. The closer the rodents lived to the city center, the worse their lung disease symptoms were.

The data posed new questions about urban squirrel populations, and the research is setting out to answer them—follow-up research is planned to determine whether the squirrel's cognition or behavior is affected by air pollution.

"The crucial point is that animals—particularly squirrels—have something to tell us. They are giving us another warning about the dangers of air pollution," said researcher Patricia Brekke of the Zoological Society of London. "They are not so much the canary in the coal mine as the rodent in the undergrowth."

Do we have the funding and appetite for more squirrel research, or will we rely on studies derived from recovery instead of following Rosalie Barrow Edge's advice to protect species while they are still common? What are our next steps in considering squirrels?

*Chapter Fifteen*

# In the Future

> "The time to protect a species is while it is still common."
> —Rosalie Barrow Edge

Every January 21, National Squirrel Appreciation Day is observed in the United States. Social media feeds fill with posts of all kinds of squirrels—common and rare, white, red, tree, ground, and even cartoon squirrels. Land conservancies, environmental organizations, and even squirrel-exasperated birders post about the celebration. Schools highlight their campus squirrel celebrities, and students post squirrel artwork.

In a twist on Groundhog Day, for over a decade February 2 has been celebrated as White Squirrel Day in Brevard, North Carolina. It's a day when Pisgah Penny takes on her famous rodent cousin Punxsutawney Phil in predicting the weather and also takes on the prediction of the winning Super Bowl team. Brevard's famous white squirrels are actually a color variant of native eastern gray squirrels, and the holiday celebration promotes their conservation.

In addition to celebrations across the United States, there are celebrations worldwide. Although the day originated in the United States in 2001, Scotland has celebrated Red Squirrel Appreciation Day on the same date since 2020.

News outlets spread the news about squirrels. The *Hindustan Times* in New Delhi, India, shared the holiday by including appropriate messages, sort of like Valentines, for readers to share: "Warm wishes on Squirrel Appreciation Day to you. Let us make it a special day for the cute squirrels out there by treating them with their favourite food"; "Squirrel Appreciation Day reminds us that there is something new to learn about these creatures which have almost 300 different species. Happy Squirrel Appreciation Day"; "Without squirrels, our gardens would be so incomplete. Let us love them and give them some extra food on Squirrel Appreciation Day. Warm wishes on this special day"; "Wishing a very Happy Squirrel Appreciation Day. Though they are so small but they are also so cute and harmless. Let us protect them and give them food."

Wherever it is feted, the purpose of the day is to show love and appreciation for local native squirrels around the world. The celebration was created by Christy Hargrove, now Christy McKeown, when she was a freshman at the University of North Carolina Asheville and a wildlife rehabilitator affiliated with the Western North Carolina Nature Center. As founder of the former *Squirrels R Us* website, Christy wrote that people could simply "celebrate by putting out extra food for the squirrels."

But Hargrove wasn't the first to suggest a specific day to commemorate the squirrels. Henry David Thoreau suggested it over a century earlier in his essay in *Faith in a Seed*:

> But what is the character of our gratitude to these squirrels—to say nothing of the others—these planters of forests, these exported dukes of Athol of many generations, which have found out how high the oak will grow on many a mountain, how low in many a valley, and how far and wide on all our plains? Are they on our pension list? Have we in any way recognized their services? We regard them as vermin. The farmer knows only that they get his seed corn occasionally in the fields adjacent to his woodlot, and perchance encourages his boys to shoot them every May, furnishing powder and shot for this purpose, while perhaps they are planting the nobler oak-corn (acorn) in its place—while up-country they have squirrel hunts on a large scale every fall and kill many thousands in a few hours, and all the neighborhood rejoices. We should be more civilized as well as humane if we recognized once in a year by some symbolical ceremony the part which the squirrel plays in the economy of Nature.

And while those celebrating today raise their voices on behalf of squirrels or put out extra food for them in their own backyards, others are following in Hargrove's other role as rehabilitators by helping one squirrel at a time.

Similar to the story of the young girl on the beach making a difference for one starfish at a time, adapted from Loren Eiseley's 1968 book *The Unexpected Universe*, these rehabilitators are saving the lives of local squirrels. Pam Spragins, another North Carolina rehabilitator and volunteer, established www.squirrel-rehab.org in 1997 to provide help for people who find squirrels in need. Spragins has been a rehabilitator, licensed by North Carolina and the US Fish and Wildlife Service, since 1991. She focuses on saving the lives of her local eastern gray squirrels, southern flying squirrels, and opossums.

Although Hargrove established a day to appreciate our bushy-tails, she wasn't alone in elevating squirrels out of the white noise each year. John Kelly, a reporter for *The Washington Post*, annually wrote Squirrel Week articles and posts. Kelly and Richard Thorington, coauthor of *Squirrels: The Animal Answer Guide*, spoke together online in 2011 about the first-ever Washington, DC, Squirrel Week. Kelly had received so many responses to his first column, in March 2011, that he "suspected there was a deep, untapped reservoir of squirrel love (and hate) out there." His Squirrel Week columns went on for years, culminating in an annual Squirrel Week photography contest providing readers with an annual array of squirrels captured by photographers from all over the country in all sorts of places, from peeking out from the hollows of trees to walking along chain-link fences to munching in backyard bird feeders. Squirrels were photographed in all environments, from western canyons and deserts to the Washington, DC, National Mall urban park. Photographers captured squirrels facing off against red-tailed hawks and barred owls. Photos shared the "deep reservoir of squirrel love (and hate)" that Kelly had recognized in the responses he had received for his column. His column and the photos lauded how we see and feel about the ubiquitous creatures that punctuate our lives with all our human emotions. Few people have been able to pull squirrels out of the background as well as John Kelly.

Have these annual squirrel revelries provided us with enough tools to move forward in valuing our common bushy-tails? Although we now know so much about squirrels—from why they crossed streets and rivers in 1968 to how they are considered keystone species, there is still so much to understand and address.

While we've spent centuries taking squirrel abundance of many species for granted, our human carelessness with those ubiquitous species has already taken a toll, as we've seen in Europe. Are Squirrel Appreciation Day and Squirrel Week enough for us to spark new studies and funding opportunities, or turn the tide on the future of the threatened Mohave ground squirrel and the northern Idaho ground squirrel, or do the celebrations just encourage more squirrel-hunting contests, which are still popular in many states?

Looking toward hunters for population data in North America, the Wild Harvest Initiative database lists four distinct squirrel species, Abert's squirrel (*Sciurus aberti*), the eastern fox squirrel (*Sciurus niger*), the eastern gray squirrel (*Sciurus carolinensis*), and the American red squirrel (*Tamiasciurus hudsonicus*), that were harvested between 2014 and 2016 in twenty-eight American states and six Canadian provinces. The total number of squirrels harvested during those two years was 16,162,719. While this provides data, that number does not reflect the impact on squirrel ecosystems or populations. And, as Emily Renn of Arizona's Habitat Harmony said about prairie dog reports, voluntary reporting by hunters likely underrepresents the numbers.

As I wrote this, the annual Big Squirrel Challenge took place in Arkansas, drawing a record 139 teams statewide at twelve sites in January 2024. While the Arkansas Game and Fish Commission website listed the winners of the contest, it did not list the totals of squirrel species harvested. After reading article after article, with headlines like "Record Participation in Big Squirrel Challenge Defies Arctic Blast" and "Big Squirrel Challenge a Hit with Hunters," announcing the contest's success with a record number of participants, I reached out to Eric Maynard, assistant chief of education for the state agency's Outdoor Skills Program, to obtain the data for the event. "We don't keep track of the number of squirrels taken," he responded. "Some teams turn in the three-squirrel limit, some turn in less."

The contest is followed by an annual World Champion Squirrel Cook Off with the theme "Squirrel—It's What's for Dinner." Drawing chefs, visitors, hunters, and television crews, the cook-off offers "organic tree-to-table squirrel" dishes that range from pizzas to ice cream flavored with local squirrel meat.

While similar contests continue in many states, a New Hampshire state senate bill was threatening to delist squirrels as game species, allowing for the indiscriminate taking of squirrels throughout the state all year long. "If Senate Bill 548 passes it enables gray squirrels to be no longer listed as game animals, it means they have zero protections. New Hampshire does not outlaw wildlife killing contests, and soon squirrel killing contests will be happening. Mother gray squirrels could be shot on sight by any New Hampshire resident, leaving dependent babies to die a horrible slow death," wrote Kristina Snyder of Chester, New Hampshire, in a letter to the editor of the *Eagle Times* on January 2, 2024. "That would not be a New Hampshire I recognize any longer. I want a state that keeps with the pulse of the people, moving towards compassion for wildlife and trends towards coexistence. I sincerely hope Senator Ward breaks the spell cast over her by her squirrel hating colleagues and finds her compassionate self again. I urge everyone to fight against Senate Bill 548." ap

If not the government agencies, who has responsibility to consider our ubiquitous wildlife, such as many squirrel species? According to Oglala Sioux holy man Black Elk, God, the Great Spirit, is within all things. "We should understand well that all things are the works of the Great Spirit. We should know that He is within all things: the trees, the grasses, the rivers, the mountains, and all the four-legged animals, and the winged peoples; and even more important, we should understand that He is also above these things and peoples." Black Elk's ability to embrace Christianity while adhering to tribal customs enabled his teachings to spread beyond his own people. His words are not unlike those of the Roman Catholic saint Francis of Assisi, who embraced all of God's creatures as his brothers and sisters. These creatures, even the small, seemingly ever-present squirrel, are included in these teachings and are believed to require our concern and care.

It was a frigid 9°F with a plummeting wind chill at my home on National Squirrel Appreciation Day 2024. Heading out into the wilds to enjoy my local squirrel population felt more than challenging. Instead, I sat bundled up in my old, leaky house, my laptop warming my lap, scrolling through the celebratory posts on social media as I put fingers to keyboard.

Months later, a new family moved in to the farm next door, with only three cats instead of the more than fifteen that had roamed throughout my own property for many years. It led me to witness the return of squirrel and chipmunk populations to my own wild yard, albeit not the reds. Of course, I treasured my spruces too much to acquiesce to my neighbor's request that I remove a healthy one near our property border, but I did see a number of his spruces removed for fear they would eventually fall. The wild, scampering squirrels are not crossing the minds of many.

Reminiscent of Ben Franklin's famous pet squirrel, a story featuring a squirrel that needed rescue and ended up as a Florida woman's pet, "I Rescued a Squirrel. Now He Refuses to Leave," ran in roughly a dozen news outlets. The adorable squirrel and its story were indeed fun. However, they also serve as a reminder of our occasional misunderstanding of wild squirrels and our well-intentioned but misguided interventions. As much as we might be tempted to keep a charming squirrel as a pet, it's important to remember that squirrels are wild animals. They deserve to remain wild, to be studied, and to thrive in their forest ecosystems. In cases when squirrels need rescue, we can take comfort in the fact that there are wildlife rehabilitators who can provide the care and expertise these animals need, rather than keeping them as pets.

Recognizing the importance of squirrels to their ecosystems is one thing; another is remembering how important they are to humans. It's not that their survival depends on their being essential to us, but that certainly does help when it comes to conservation and recovery funds.

While we know all the reasons why squirrels can benefit their native ecosystems and can entertain us, are you prepared for one job that you probably wouldn't think they'd be suited for—how about sniffer squirrels? Although dogs have a vast number of olfactory receptors in their nose and their own data processing center in their brain to be successful at detection, squirrels have a more extensive nasal epithelium. Like dogs, they also use their sense of smell to locate food and detect threats. So why not try them in the role of sniffer squirrels? The Public Security Bureau of the Hechuan District in Chongqing, China, trained six Eurasian red squirrels to sniff out drugs, and they are hailed as being equally efficient as sniffer dogs. Squirrels alert their handlers to finding a narcotic substance by scratching, something they do much better than barking. Their small size enables them to enter places where dogs might not fit. Police dog trainer Yin Jin told the *Chongqing Daily* that it had taken years to get the squirrels to this ability level, but they were doing an "excellent job." The squirrels follow in the footsteps of rats used to detect land mines in Cambodia. Challenging the skills of squirrels has been untested up until now.

While used for detection, like sniffer dogs, squirrels have also served as models for human functions, pathology, and disease. For example, gray squirrels have been used to study human lenses and color vision, fox squirrels for metabolic disorders.

Our enthusiasm and attitude toward squirrels will continue to ebb and flow. Our relationship with Central Park's squirrels is an example of this ongoing complex, intermittent relationship. First welcomed, Central Park squirrels were hunted in the 1880s after the population gained numbers. One city hunt took out hundreds of squirrels in the park on a single morning. However, a decade later, in the 1890s, hunting of squirrels became unpopular again when Italian laborers were accused of hunting squirrels for food. Periods of feeding park squirrels alternated with periods of not feeding and periods of hunting. Were the squirrels overfed or underfed? Were they harming the bird population or biting people? Enthusiasm waned and waxed. Over a century later, in 2010, the New York City Department of Parks and Recreation embraced the city's urban squirrel populations, gray, fox, and flying, by naming Pearl the Squirrel its official mascot. But Pearl became the perfect example of our intermittent relationship with squirrels when the mascot was "fired from the field" just two years later. Park spokesman Sam Biederman said to the *New York Post* in 2017, "She has since taken a desk job in the communications department."

In other parts of the country, we've seen centuries of competitive squirrel hunts, such as the recent hunt in New York's North Country, and at the same time arrests throughout the same state for squirrel killing. In 2010, for example, an eighty-one-year-old US Army veteran was sentenced to thirty days in prison for shooting a squirrel in his Long Island yard. A man was arrested in the lower Hudson Valley in 2024 for trapping and spray-painting the squirrels in his yard with Apple Red Rust-Oleum paint after they allegedly riled up his dog. Attitudes and laws continue to shift and change.

While there are those humans fixated on ridding their yards of squirrels, the complaints are often without merit. Squirrels, for example, rarely transmit rabies—no need for the extra worry. The man arrested for spray-painting the squirrels might have been more concerned about his local squirrels transmitting fleas or ticks to his dogs than just riling them up.

The very nature of squirrels has clashed with the essence of humanity throughout history. Do we have the luxury of allowing this ebb and flow of attention to squirrels in this time of increased habitat fragmentation and global climate change?

Ernest Thompson Seton contemplated the choice between eradicating squirrels for their marauding behavior and fostering their population in his 1909 book, *Life-Histories of Northern Animals*:

> Should we preserve the Red-squirrels in view of the fact that they destroy a certain amount of grain, fruit, and song-birds every year? These are serious charges, and I cannot refute them in detail; but I know that my grounds abound now, as they have for years, with grain, fruit, song-birds, and Red-squirrels, showing that these are not incompatible. They are near some sort of balance. It may prove a wise thing to keep the Chickaree numbers down since their natural foe, the Marten, is gone from New England, but I am far from joining with those who would welcome its extinction. Indeed, I should woefully miss the noisy little rascals if I did not see them at their daily play, and I hope that the Red-squirrels will frequent my grounds at least as long as I do.

There are things we can do individually to create more data about our squirrel populations. For example, there is a participatory science initiative in New England to uncover data about flying squirrels, both northern and southern species. The New England Flying Squirrel Network was established to gather data that would provide a better concept of the population of these species. Berkshire Community College professor Tom Tyning remarked, "My experience is that if you tap on about 1,000 trees with abandoned woodpecker holes, one of them will have a flying squirrel in it!" The network is looking for backyard naturalists to install a flying squirrel nest box on a tree that can be monitored once a month. While participants are encouraged not to handle any squirrels or other wildlife, they can provide important data through their observations. Catherine J. Wong, assistant professor and engineering and physical sciences librarian at the University of New Hampshire, has made it a mission to get others interested in putting up nest boxes to participate in this project.

Roads continue to dissect our communities and fragment our ecosystems. A mere fifty or so years ago, just 3 percent of land-dwelling mammals met their end on a highway, according to Ben Goldfarb in his book *Crossings*, but that percentage had quadrupled by 2017.

However, there can be change. We witness that each time a crossing is constructed to lessen wildlife losses. In India, Squirrel Appreciation Day is even more significant. It is where the Kerala Forest Department is building canopy bridges to ensure the threatened Indian giant squirrels can avoid cars as they move through their habitat. A canopy bridge, created from dried twigs, bamboo, wild creepers, and rope, was built at the Chinnar Wildlife Sanctuary and the Marayoor-Udumalpetta Road for the grizzled giant squirrel. But those squirrel bridges are rare among squirrel protections across the world.

Returning to the State of the State of the squirrels in our world, we find that humans are continuing their age-old love-hate relationship with the small, bushy-tailed rodents; habitats are becoming more fragmented; invasive nonnative squirrels are endangering native squirrel populations; and we don't accurately know the status of roughly 40 percent of the world's squirrel species. But the stories continue in our books, on our screens, and even in our video games, with the launch of the game *Squirrel with a Gun*. The new game includes players cast as diminutive armed squirrels with ambitions exceeding their stature, carrying an array of firearms. According to *Hardcore Gamer*, the game humorously juxtaposes the natural world with the surreal scenario of an armed squirrel. Players must navigate their squirrel protagonists through challenges, evading agents working earnestly to halt their chaotic spree.

Studies of how squirrels and other wildlife interact with humans in cities are increasingly important as urban centers grow and humans move into wildlife habitats. Additionally, it is vital to increase our grasp of the impact of pollution on wildlife and humans.

Researchers from Britain's University of Chester and Japan's Hokkaido University studied Eurasian red squirrel behavior, problem-solving, and cognition in eleven urban centers in Obihiro, Japan. They discovered that urban environmental characteristics are stressors for squirrels and can shape cognitive performance. Ultimately, their research led them to determine that the relationship between urban environments and wildlife cognition is mainly unclear and requires more investigation.

It is time to consider our global squirrel species—their portrayal, how they enrich our world, and how we study them. We must consider them when managing our forests and in the way we coexist with them and other wildlife. We must consider whether our ubiquitous squirrel species are genuinely renewable resources. In considering our many squirrel species, we must also deliberately protect our planet so that even our squirrels do not suffer the plight of other formerly abundant wildlife we've already lost.

"Squirrels help us study how ecosystems are changing, and the recovery from these changes is critical," said John Koprowski. "The challenges involve humans as well as natural systems. We can help effect change with management elements. It's an exciting time to be a wildlife biologist." When he gave the opening address at the Seventh International Colloquium on Arboreal Squirrels in Helsinki, Finland, he asked: "Listen to the chatter: what can squirrels teach us about conservation in a changing world?"

While considering the squirrel and its many worldwide species and their survival challenges, let's not neglect to reflect on and celebrate the good news. A 2014 study by Emiliano Mori of the Italian National Research Council and Mattia Menchetti of the Institute of Evolutionary Biology revealed that there is sometimes good news in a country where most news involves the challenges in the face of an invasive species. With the help of citizen science, the red squirrel, facing habitat fragmentation and competition from invasive grays, underwent a range expansion in southern Latium since 2005. The squirrel subspecies, *Sciurus vulgaris italicus*, was discovered in northern Campania, on the border with Latium and Molise. Of course, it required further research.

This subspecies of the Eurasian red is one of three: *Sciurus vulgaris fuscoater*, found in northern Italy in the Alps and northern Apennines; *Sciurus vulgaris italicus*, found in Central Italy, which is a bit smaller, has a range of coat colors, and was presumed extinct; and *Sciurus vulgaris meridionalis* Lucifero, 1907, the most prominent Italian subspecies, endemic to the southern Apennines. Scientists use skull measurements to identify the subspecies.

Across the world, a student researcher from the University of Illinois captured the image of an elusive native southern flying squirrel using a camera trap. Wildlife ecologists say it was the first one documented in the area in decades. We hope that scientists will continue unlocking the mysteries involving squirrel species and possibly discovering new and rare squirrels.

As we celebrate these hopeful stories of squirrel comebacks, we can also consider how we introduce children to ubiquitous wildlife. If they are becoming excited solely by seeing exotic wildlife in our zoos, how will they become stewards of familiar, pedestrian wildlife? Zoos connect children to animals, but, as Etienne Benson reminds us, which animals? "What does it do to children if they feel that the special place [the zoo] where they see animals are the ones they never see in their daily lives?" After all, we must consider Rosalie Barrow Edge's philosophy of protecting species while they are still common. But in order to protect them, we must first consider them.

## *Acknowledgments*

In many ways, this book began when, as a toddler, I fed a peanut to a chipmunk on the porch of a Lake George log cabin in New York's Adirondack Park. It is my earliest memory of interacting with a wild mammal, and it was an activity repeated for many more summers. Perhaps that friendly transactional relationship, jump-started by my nature-loving parents, spurred my lifelong passion for wild things. I have to begin by thanking them for teaching me, from the start, to love, appreciate, and protect the natural world, from its rocks to plants to animals.

Many individuals and organizations helped me in the course of writing *Squirrel*, and I'd like to acknowledge as many as I can. Writing this book entailed abridgment, and I beg forgiveness from those people and entities I have inadvertently left out. *Squirrel* is anything but the final volume on our planet's squirrel species, and I fervently hope that it prompts its readers to explore further and consider our many squirrels, the seemingly ubiquitous creatures we tend to overlook.

Libraries are the lifeline for research. My thanks to the librarians at the University at Albany Science Library, the New York Public Library, the New York State Library, the Library of Congress, and the Historical Society of Pennsylvania, among others. In addition, my gratitude extends to the US Fish and Wildlife Service staff who responded to my Freedom of Information Act (FOIA) requests.

During this reporting, I spoke and interacted with numerous experts—some named on these pages. Each of these professionals, named and not, contributed to my understanding. Dr. Etienne Benson, Collette Yee, Dr. Noah Perlut and his students Autumn Linden and Rick Hamilton, Francesca Sironi, Kelly Wallis, Elise Wiggins, Dr. Maria Mazzamuto, Dr. Sandro Bertolino, Dr. Tara Harris, Jessica Simmons, Jeff Gagnon, Adam Rose, and Emily Renn were notably generous with their time.

My editor, Erin Johnson, saw the potential in this book, championed it, and gave it a worthy home. I am forever thankful. Many thanks to everyone at Island Press who joined Team Squirrel. For more than a decade, the support from my agent, Jennifer Laughran, has helped me add books to my shelf and friendship to my days. Early readers of my work in progress are always vital; I have some of the best. Many thanks to Anita Sanchez and Lois Huey, among others. And those friends and family who filled my life and office with squirrel pins, earrings, trinkets, and more to add smiles during this fun project—Lucie Castaldo, Carrie Pearson, Amy Traggianese, and Nancy Miles. Finally, I am more than grateful to my husband, Dean, for being my partner in life, love, and the pursuit of squirrel adventures.

## Coexisting with Squirrels

Here are some tips to help you be a good neighbor to your native wild squirrels.

1. Provide water in a birdbath or shallow dish. Keep it clean.
2. Keep cats indoors so that they won't hunt your native wildlife.
3. Install a feeder during cold weather months. Fill it with tree nuts in the shell, such as walnuts, hickory nuts, and acorns. But remember that feeding wildlife interrupts their natural behavior and can alter their natural diet. Be consistent.
4. Plant native trees, both deciduous species and conifers, for safety, food, and nesting.
5. Plant native shrubs, such as native wild cherry, that provide food and shelter for squirrels and other wildlife.
6. Keep wild animals wild. Although it is tempting to interact with them, don't allow them to lose their natural fear of people.
7. Be aware of local laws where you live and travel, such as Costa Rica's Law of Conservation of Wildlife No. 7317, which prohibits supplying food to wildlife in that country.
8. If you come across a young squirrel, consider that a newborn's highest probability for survival is with its mother, so give the mother time to return to her little one. If she doesn't return, take the young squirrel to a wildlife rehabilitator. Do not consider keeping it as a pet.

### Follow the 5-7-9 Rule to Keep Squirrels from Your Bird Feeder

Follow this popular rule that provides you with a squirrel-free safety zone for your feeder. It takes into account how far a tree squirrel can jump. Install your feeder 5 feet off the ground, 7 feet from your house or trees, and at least 9 feet from anything hanging above it.

*Sources:* Janet Connolly, "How to Keep Squirrels Off of Feeders Forever," The Audubon Shop, July 9, 2020, https://www.theaudubonshop.com/how-to-keep-squirrels-off-of-feeders-forever/; Enviroliteracy Team, "What Is the 5-7-9 Rule for Bird Feeders?," Environmental Literacy Council, April 8, 2025, https://enviroliteracy.org/what-is-the-5-7-9-rule-for-bird-feeders/.

# *Squirrel Family Tree Classification*

Family Sciuridae (squirrels)
Subfamily Callosciurinae (Asian squirrels)
Subfamily Xerinae (marmots, ground squirrels, African squirrels, and relatives)
Subfamily Sciurinae (tree squirrels, flying squirrels, and relatives)
Subfamily Ratufinae (Asian giant squirrels)
Subfamily Sciurillinae (pygmy squirrels)

**World Squirrel Species (IUCN Red List of Threatened Species Status, 2025)**
Note: These are global species with global status notes. Species might have a differing local or federal status.

Within subcategories, species listed here are grouped by genus and then alphabetized by common name.

Subfamily Callosciurinae (Asian squirrels)
Anderson's squirrel (*Callosciurus quinquestriatus*): Unknown
Black-striped squirrel (*Callosciurus nigrovittatus*): Stable
Borneo black-banded squirrel (*Callosciurus orestes*): Stable
Ear-spot squirrel (*Callosciurus adamsi*): Decreasing/Near Threatened
Finlayson's squirrel (*Callosciurus finlaysonii*): Stable
Grey-bellied squirrel (*Callosciurus caniceps*): Stable
Hoary-bellied squirrel (*Callosciurus pygerythrus*): Stable
Inornate squirrel (*Callosciurus inornatus*): Stable
Kinabalu squirrel (*Callosciurus baluensis*): Decreasing
Kloss's squirrel (*Callosciurus albescens*): Unknown
Mentawai squirrel (*Callosciurus melanogaster*): Decreasing/Vulnerable
Pallas's squirrel (*Callosciurus erythraeus*): Stable
Phayre's squirrel (*Callosciurus phayrei*): Unknown
Plantain squirrel (*Callosciurus notatus*): Increasing
Prevost's squirrel (*Callosciurus prevostii*): Decreasing

Subfamily Xerinae
Tribe Marmotini
Alaska marmot (*Marmota broweri*): Stable
Alpine marmot (*Marmota marmota*): Unknown
Altai marmot (*Marmota baibacina*): Unknown
Black-capped marmot (*Marmota camtschatica*): Unknown
Bobak marmot (*Marmota bobak*): Stable
Groundhog (*Marmota monax*): Stable

Himalayan marmot (*Marmota himalayana*): Unknown
Hoary marmot (*Marmota caligata*): Stable
Long-tailed marmot (*Marmota caudata*): Unknown
Menzbier's marmot (*Marmota menzbieri*): Decreasing/Vulnerable/
 Critically Depleted
Mongolian marmot (*Marmota sibirica*): Decreasing/Endangered
Olympic marmot (*Marmota olympus*): Decreasing
Vancouver Island marmot (*Marmota vancouverensis*): Decreasing/
 Critically Endangered (90 adults remaining)
Yellow-bellied marmot (*Marmota flaviventris*): Stable

Prairie Dogs
Black-tailed prairie dog (*Cynomys ludovicianus*): Decreasing
Gunnison's prairie dog (*Cynomys gunnisoni*): Decreasing
Mexican prairie dog (*Cynomys mexicanus*): Decreasing/Endangered
Utah prairie dog (*Cynomys parvidens*): Decreasing/Endangered
White-tailed prairie dog (*Cynomys leucurus*): Decreasing

Ground Squirrels
Barbary ground squirrel (*Atlantoxerus getulus*): Stable

Cascade golden-mantled ground squirrel (*Callospermophilus saturatus*): Stable
Golden-mantled ground squirrel (*Callospermophilus lateralis*): Stable
Sierra Madre ground squirrel (*Callospermophilus madrensis*): Unknown/
 Near Threatened

Bornean mountain ground squirrel (*Dremomys everetti*): Stable

Mexican ground squirrel (*Ictidomys mexicanus*): Stable
Thirteen-lined ground squirrel (*Ictidomys tridecemlineatus*): Stable

Niobe ground squirrel (*Lariscus niobe*): Unknown

Ring-tailed ground squirrel (*Notocitellus annulatus*): Unknown
Tropical ground squirrel (*Notocitellus adocetus*): Stable

California ground squirrel (*Otospermophilus beecheyi*): Stable

Franklin's ground squirrel (*Poliocitellus franklinii*): Decreasing

Tufted ground squirrel (*Rheithrosciurus macrotis*): Decreasing/Vulnerable

Alashan ground squirrel (*Spermophilus alashanicus*): Decreasing
Anatolian ground squirrel (*Spermophilus xanthoprymnus*): Decreasing/
 Near Threatened
Brandt's ground squirrel (*Spermophilus brevicauda*): Unknown
Caucasian Mountain ground squirrel (*Spermophilus musicus*): Increasing
Daurian ground squirrel (*Spermophilus dauricus*): Unknown
European ground squirrel (*Spermophilus citellus*): Decreasing/Endangered
Little ground squirrel (*Spermophilus pygmaeus*): Decreasing
Long-clawed ground squirrel (*Spermophilopsis leptodactylus*): Unknown

Pallid ground squirrel (*Spermophilus pallidicauda*): Unknown
Red-cheeked ground squirrel (*Spermophilus erythrogenys*): Stable
Relict ground squirrel (*Spermophilus relictus*): Unknown
Russet ground squirrel (*Spermophilus major*): Decreasing/Near Threatened
Speckled ground squirrel (*Spermophilus suslicus*): Decreasing/Critically Endangered
Taurus ground squirrel (*Spermophilus taurensis*): Unknown
Tien Shan ground squirrel (*Spermophilus ralli*): Unknown
Yellow ground squirrel (*Spermophilus fulvus*): Unknown

Merriam's ground squirrel (*Urocitellus canus*): Unknown
Northern Idaho ground squirrel (*Urocitellus brunneus*): Increasing/Endangered
Piute ground squirrel (*Urocitellus mollis*): Stable
Richardson's ground squirrel (*Urocitellus richardsonii*): Stable
Southern Idaho ground squirrel (*Urocitellus endemicus*): Unknown/Vulnerable
Townsend's ground squirrel (*Urocitellus townsendii*): Decreasing/Vulnerable
Uinta ground squirrel (*Urocitellus armatus*): Unknown
Washington ground squirrel (*Urocitellus washingtoni*): Decreasing/Near Threatened
Wyoming ground squirrel (*Urocitellus elegans*): Unknown

Mohave ground squirrel (*Xerospermophilus mohavensis*): Decreasing/Near Threatened
Perote ground squirrel (*Xerospermophilus perotensis*): Decreasing/Endangered
Round-tailed ground squirrel (*Xerospermophilus tereticaudus*): Stable
Spotted ground squirrel (*Xerospermophilus spilosoma*): Stable

Damara ground squirrel (*Xerus princeps*): Stable
South African ground squirrel (*Xerus inauris*): Stable
Striped ground squirrel (*Xerus erythropus*): Stable
Unstriped ground squirrel (*Xerus rutilus*): Stable

Cameroon scaly-tail (*Zenkerella insignis*): Unknown

Chipmunks
Siberian chipmunk (*Eutamias sibiricus*): Stable

Alpine chipmunk (*Neotamias alpinus*): Stable
Buller's chipmunk (*Neotamias bulleri*): Decreasing/Vulnerable
California chipmunk (*Neotamias obscurus*): Stable
Cliff chipmunk (*Neotamias dorsalis*): Stable
Colorado chipmunk (*Neotamias quadrivittatus*): Stable
Durango chipmunk (*Neotamias durangae*): Unknown
Gray-collared chipmunk (*Neotamias cinereicollis*): Stable
Gray-footed chipmunk (*Neotamias canipes*): Stable
Hopi chipmunk (*Neotamias rufus*): Stable
Least chipmunk (*Neotamias minimus*): Stable
Lodgepole chipmunk (*Neotamias speciosus*): Stable
Long-eared chipmunk (*Neotamias quadrimaculatus*): Stable
Merriam's chipmunk (*Neotamias merriami*): Stable
Palmer's chipmunk (*Neotamias palmeri*): Decreasing/Endangered

Panamint chipmunk (*Neotamias panamintinus*): Stable
Red-tailed chipmunk (*Neotamias ruficaudus*): Stable
Shadow chipmunk (*Neotamias senex*): Stable
Siskiyou chipmunk (*Neotamias siskiyou*): Stable
Sonoma chipmunk (*Neotamias sonomae*): Stable
Townsend's chipmunk (*Neotamias townsendii*): Stable
Uinta chipmunk (*Neotamias umbrinus*): Stable
Yellow-cheeked chipmunk (*Neotamias ochrogenys*): Stable
Yellow-pine chipmunk (*Neotamias amoenus*): Stable

Eastern chipmunk (*Tamias striatus*): Stable

**Subfamily Sciurinae**
Flying Squirrels—Tribe Pteromyini
Northern Chinese flying squirrel (*Aeretes melanopterus*): Stable

Black flying squirrel (*Aeromys tephromelas*): Unknown
Thomas's flying squirrel (*Aeromys thomasi*): Unknown

Hairy-footed flying squirrel (*Belomys pearsonii*): Unknown

Laotian giant flying squirrel (*Biswamoyopterus laoensis*): Unknown
Namdapha flying squirrel (*Biswamoyopterus biswasi*): Decreasing/Critically Endangered

Small Kashmir flying squirrel (*Eoglaucomys fimbriatus*): Unknown

Woolly flying squirrel (*Eupetaurus cinereus*): Decreasing/Endangered

Northern flying squirrel (*Glaucomys sabrinus*): Stable
Southern flying squirrel (*Glaucomys volans*): Stable

Arrow flying squirrel (*Hylopetes sagitta*): Unknown
Bartels's flying squirrel (*Hylopetes bartelsi*): Unknown
Indochinese flying squirrel (*Hylopetes phayrei*): Stable
Jentink's flying squirrel (*Hylopetes platyurus*): Unknown
Palawan flying squirrel (*Hylopetes nigripes*): Decreasing/Near Threatened
Particolored flying squirrel (*Hylopetes alboniger*): Decreasing
Red-cheeked flying squirrel (*Hylopetes spadiceus*): Unknown
Sipora flying squirrel (*Hylopetes sipora*): Decreasing/Endangered
Sumatran flying squirrel (*Hylopetes winstoni*): Unknown

Long-eared scaly-tailed flying squirrel (*Idiurus macrotis*): Unknown
Pygmy scaly-tailed flying squirrel (*Idiurus zenkeri*): Unknown

Javanese flying squirrel (*Iomys horsfieldii*): Stable
Mentawi flying squirrel (*Iomys sipora*): Decreasing/Endangered

Hose's pygmy flying squirrel (*Petaurillus hosei*): Unknown
Lesser pygmy flying squirrel (*Petaurillus emiliae*): Unknown
Selangor pygmy flying squirrel (*Petaurillus kinlochii*): Unknown

Bhutan giant flying squirrel (*Petaurista nobilis*): Decreasing/Near Threatened
Chinese giant flying squirrel (*Petaurista xanthotis*): Unknown
Hodgson's giant flying squirrel (*Petaurista magnificus*): Decreasing
Indian giant flying squirrel (*Petaurista philippensis*): Decreasing
Japanese giant flying squirrel (*Petaurista leucogenys*): Unknown
Mechuka giant flying squirrel (*Petaurista mechukaensis*): Unknown/Near Threatened
Mishmi giant flying squirrel (*Petaurista mishmiensis*): Unknown/Near Threatened
Red and white giant flying squirrel (*Petaurista alborufus*): Unknown
Red giant flying squirrel (*Petaurista petaurista*): Decreasing
Spotted giant flying squirrel (*Petaurista elegans*): Stable

Basilan flying squirrel (*Petinomys crinitus*): Stable
Hagen's flying squirrel (*Petinomys hageni*): Decreasing
Mindanao flying squirrel (*Petinomys mindanensis*): Unknown
Siberut flying squirrel (*Petinomys lugens*): Decreasing/Vulnerable
Temminck's flying squirrel (*Petinomys setosus*): Decreasing/Vulnerable
Travancore flying squirrel (*Petinomys fuscocapillus*): Decreasing
Vordermann's flying squirrel (*Petinomys vordermanni*): Decreasing/Vulnerable
Whiskered flying squirrel (*Petinomys genibarbis*): Decreasing/Vulnerable

Japanese flying squirrel (*Pteromys momonga*): Unknown
Siberian flying squirrel (*Pteromys volans*): Decreasing

Smoky flying squirrel (*Pteromyscus pulverulentus*): Decreasing/Endangered

Complex-toothed flying squirrel (*Trogopterus xanthipes*): Decreasing/Near Threatened
Tree Squirrels, Red Squirrels, and Relatives—Tribe Sciurini
Harris's antelope squirrel (*Ammospermophilus harrisii*): Unknown
Nelson's antelope squirrel (*Ammospermophilus nelsoni*): Decreasing/Endangered
Texas antelope squirrel (*Ammospermophilus interpres*): Unknown
White-tailed antelope squirrel (*Ammospermophilus leucurus*): Stable

Asian red-cheeked squirrel (*Dremomys rufigenis*): Stable
Orange-bellied Himalayan squirrel (*Dremomys lokriah*): Decreasing
Perny's long-nosed squirrel (*Dremomys pernyi*): Stable
Red-hipped squirrel (*Dremomys pyrrhomerus*): Unknown
Red-throated squirrel (*Dremomys gularis*): Unknown

Western palm squirrel (*Epixerus ebii*): Unknown

Common palm squirrel (*Funambulus palmarum*): Increasing
Dusky striped squirrel (*Funambulus obscurus*): Decreasing/Vulnerable
Dusky-striped squirrel (*Funambulus sublineatus*): Decreasing/Vulnerable
Five-striped palm squirrel (*Funambulus pennantii*): Unknown
Layard's palm squirrel (*Funambulus layardi*): Decreasing/Vulnerable
Western Ghats striped squirrel (*Funambulus tristriatus*): Decreasing

Congo rope squirrel (*Funisciurus congicus*): Stable
Du Chaillu's rope squirrel (*Funisciurus duchaillui*): Unknown

Fire-footed rope squirrel (*Funisciurus pyrropus*): Stable
Kintampo rope squirrel (*Funisciurus substriatus*): Unknown
Lady Burton's rope squirrel (*Funisciurus isabella*): Unknown
Leconte's four-striped tree squirrel (*Funisciurus lemniscatus*): Unknown
Lunda rope squirrel (*Funisciurus bayonii*): Unknown
Red-cheeked rope squirrel (*Funisciurus leucogenys*): Unknown
Ribboned rope squirrel (*Funisciurus carruthersi*): Unknown
Thomas's rope squirrel (*Funisciurus anerythrus*): Unknown

Sculptor squirrel (*Glyphotes simus*): Unknown

Gambian sun squirrel (*Heliosciurus gambianus*): Unknown
Mutable sun squirrel (*Heliosciurus mutabilis*): Unknown
Red-legged sun squirrel (*Heliosciurus rufobrachium*): Unknown
Ruwenzori sun squirrel (*Heliosciurus ruwenzorii*): Unknown
Small sun squirrel (*Heliosciurus punctatus*): Unknown
Zanj sun squirrel (*Heliosciurus undulatus*): Unknown

Lowland long-nosed squirrel (*Hyosciurus ileile*): Unknown
Montane long-nosed squirrel (*Hyosciurus heinrichi*): Unknown

Four-striped ground squirrel (*Lariscus hosei*): Decreasing
Mentawai three-striped squirrel (*Lariscus obscurus*): Decreasing/Near Threatened
Three-striped ground squirrel (*Lariscus insignis*): Decreasing

Berdmore's squirrel (*Menetes berdmorei*): Stable

Alfaro's pygmy squirrel (*Microsciurus alfari*): Stable
Amazon dwarf squirrel (*Microsciurus flaviventer*): Unknown
Santander dwarf squirrel (*Microsciurus santanderensis*): Unknown
Western dwarf squirrel (*Microsciurus mimulus*): Stable

Black-eared squirrel (*Nannosciurus melanotis*): Decreasing

Rock squirrel (*Otospermophilus variegatus*): Stable

Alexander's bush squirrel (*Paraxerus alexandri*): Unknown
Black and red bush squirrel (*Paraxerus lucifer*): Unknown
Boehm's bush squirrel (*Paraxerus boehmi*): Unknown
Cooper's mountain squirrel (*Paraxerus cooperi*): Unknown
Green bush squirrel (*Paraxerus poensis*): Unknown
Ochre bush squirrel (*Paraxerus ochraceus*): Stable
Red bush squirrel (*Paraxerus palliatus*): Unknown
Smith's bush squirrel (*Paraxerus cepapi*): Stable
Striped bush squirrel (*Paraxerus flavovittis*): Unknown
Swynnerton's bush squirrel (*Paraxerus vexillarius*): Unknown
Vincent's bush squirrel (*Paraxerus vincenti*): Decreasing/Endangered

Alston's squirrel (*Prosciurillus alstoni*): Unknown/Near Threatened
Celebes dwarf squirrel (*Prosciurillus murinus*): Unknown
Mount Topapu squirrel (*Prosciurillus topapuensis*): Unknown/Near Threatened

Sanghir squirrel (*Prosciurillus rosenbergii*): Unknown/Endangered
Secretive dwarf squirrel (*Prosciurillus abstrusus*): Unknown/Near Threatened

Forest giant squirrel (*Protoxerus stangeri*): Unknown
Slender-tailed squirrel (*Protoxerus aubinnii*): Unknown/Near Threatened
Weber's dwarf squirrel (*Prosciurillus weberi*): Unknown/Vulnerable
Whitish dwarf squirrel (*Prosciurillus leucomus*): Unknown

Long-nosed or shrew-faced squirrel (*Rhinosciurus laticaudatus*): Decreasing/Near Threatened

Red-bellied or Sulawesi giant squirrel (*Rubrisciurus rubriventer*): Unknown

Neotropical pygmy squirrel (*Sciurillus pusillus*): Unknown

Abert's squirrel (*Sciurus aberti*): Stable
Allen's squirrel (*Sciurus alleni*): Decreasing
Andean squirrel (*Sciurus pucheranii*): Unknown
Arizona gray squirrel (*Sciurus arizonensis*): Unknown
Bolivian squirrel (*Sciurus ignitus*): Unknown
Calabrian black squirrel (*Sciurus meridionalis*): Stable/Near Threatened
Caucasian squirrel (*Sciurus anomalus*): Decreasing
Collie's squirrel (*Sciurus colliaei*): Unknown
Deppe's squirrel (*Sciurus deppei*): Stable
Eastern fox squirrel (*Sciurus niger*): Stable
Eastern gray squirrel (*Sciurus carolinensis*): Increasing
Eurasian red squirrel (*Sciurus vulgaris*): Unknown
Fiery squirrel (*Sciurus flammifer*): Unknown
Guayaquil squirrel (*Sciurus stramineus*): Unknown
Guianan squirrel (*Sciurus aestuans*): Unknown
Japanese squirrel (*Sciurus lis*): Stable
Junín red squirrel (*Sciurus pyrrhinus*): Unknown
Mexican fox squirrel (*Sciurus nayaritensis*): Unknown
Northern Amazonian red squirrel (*Sciurus igniventris*): Unknown
Peters's squirrel (*Sciurus oculatus*): Unknown
Red-bellied squirrel (*Sciurus aureogaster*): Stable
Red-tailed squirrel (*Sciurus granatensis*): Stable
Richmond's squirrel (*Sciurus richmondi*): Unknown/Near Threatened
Sanborn's squirrel (*Sciurus sanborni*): Unknown
Southern Amazon red squirrel (*Sciurus spadiceus*): Unknown
Variegated squirrel (*Sciurus variegatoides*): Stable
Western gray squirrel (*Sciurus griseus*): Unknown
Yellow-throated squirrel (*Sciurus gilvigularis*): Unknown
Yucatan squirrel (*Sciurus yucatanensis*): Stable

Forrest's rock squirrel (*Sciurotamias forresti*): Unknown
Pére David's rock squirrel (*Sciurotamias davidianus*): Unknown

Brooke's squirrel (*Sundasciurus brookei*): Unknown
Busuanga squirrel (*Sundasciurus hoogstraali*): Unknown

Culion tree squirrel (*Sundasciurus moellendorffi*): Decreasing/Near Threatened
Davao squirrel (*Sundasciurus davensis*): Unknown
Fraternal squirrel (*Sundasciurus fraterculus*): Decreasing/Vulnerable
Horse-tailed squirrel (*Sundasciurus hippurus*): Decreasing/Near Threatened
Jentink's squirrel (*Sundasciurus jentinki*): Unknown
Low's squirrel (*Sundasciurus lowii*): Stable
Mindanao squirrel (*Sundasciurus mindanensis*): Unknown
Northern Palawan tree squirrel (*Sundasciurus juvencus*): Stable
Palawan montane squirrel (*Sundasciurus rabori*): Unknown
Philippine tree squirrel (*Sundasciurus philippinensis*): Stable
Samar squirrel (*Sundasciurus samarensis*): Unknown
Slender squirrel (*Sundasciurus tenuis*): Decreasing
Southern Palawan tree squirrel (*Sundasciurus steerii*): Stable

Bangs's mountain squirrel (*Syntheosciurus brochus*): Decreasing

Douglas's squirrel (*Tamiasciurus douglasii*): Stable
Mearns's squirrel (*Tamiasciurus mearnsi*): Decreasing/Endangered
Red squirrel (*Tamiasciurus hudsonicus*): Stable

Cambodian striped squirrel (*Tamiops rodolphii*): Stable
Himalayan striped squirrel (*Tamiops mcclellandii*): Stable
Maritime striped squirrel (*Tamiops maritimus*): Stable
Swinhoe's striped squirrel (*Tamiops swinhoei*): Stable

**Subfamily Ratufinae (Asian giant squirrels)**
Black giant squirrel (*Ratufa bicolor*): Decreasing/Near Threatened
Indian giant squirrel (*Ratufa indica*): Decreasing
Pale giant squirrel (*Ratufa affinis*): Decreasing/Near Threatened
Sri Lankan giant squirrel (*Ratufa macroura*): Decreasing/Near Threatened

**Subfamily Sciurillinae (pygmy squirrels)**
Least pygmy squirrel (*Exilisciurus exilis*): Unknown
Philippine pygmy squirrel (*Exilisciurus concinnus*): Stable
Tufted pygmy squirrel (*Exilisciurus whiteheadi*): Stable

African pygmy squirrel (*Myosciurus pumilio*): Unknown

# Source Notes

### Epigraph

**"So far as our noblest hardwood":** Henry David Thoreau, *Faith in a Seed: The Dispersion of Seeds and Other Late Natural History Writings*, edited by Bradley Dean (Island Press, 1996), 130.

**"The last word in ignorance":** Aldo Leopold, *Round River: From the Journals of Aldo Leopold* (Oxford University Press, 1993), 146.

**"For it is the right and property":** Ralph Waldo Emerson, "Art," accessed March 28, 2023, https://emersoncentral.com/texts/essays-first-series/art/.

### Preface

**currently there are 294 identified squirrel species:** International Union for Conservation of Nature and Natural Resources (IUCN), Red List of Threatened Species, version 2025-1, accessed May 14, 2025, https://www.iucnredlist.org/search/list?query=Squirrels&searchType=species.

**"There's actually so little known":** Kate Morgan, "How (and Why) to Peacefully Coexist with Squirrels," *The Washington Post*, February 21, 2024, https://www.washingtonpost.com/home/2024/02/22/living-with-squirrels-nuisance/.

**"While unquestionably the most":** "Project Squirrel," University of New England, School of Marine and Environmental Programs, Perlut Lab, accessed February 21, 2024, https://sites.une.edu/perlutlab/project-squirrel/.

**"They've been studied pretty well":** Noah Perlut, interview by the author, October 9, 2024.

**"Some people are bird people":** Jason Daley, "Volunteers Counted All the Squirrels in Central Park," *Smithsonian Magazine*, June 24, 2019, https://www.smithsonianmag.com/smart-news/new-census-counted-all-squirrels-central-park-180972480/.

**"We don't really think":** Morgan, "How (and Why) to Peacefully Coexist with Squirrels."

**"squirrels are excellent and understudied models":** John Koprowski, "Mitä Hyötyä Oravista On?" *Kioski Vieraili Oravakonferenssissa: "Oravat Istuttavat Puita Ja Levittävät Sieniä,"* August 6, 2017, https://vimeo.com/129859604.

**"create pockets of rural peace and calm":** Etienne Benson, "The Urbanization of the Eastern Gray Squirrel in the United States," *Journal of American History* 100, no. 3 (December 2013): 691–710, https://doi.org/10.1093/jahist/jat353; Nina Strochlic,

"Squirrels Were Purposefully Introduced to American Cities," *National Geographic*, February 21, 2018, https://www.nationalgeographic.com/magazine/article/explore-city-squirrels-nuisance.

**"The time to protect a species":** Author's visit to Hawk Mountain Sanctuary, April 4, 2024.

## Chapter 1

**"Keystone: a central stone":** *Oxford English Dictionary* (Oxford University Press, 1992).

**"I learned a fun fact about squirrels":** "Sarah Silverman: A Speck of Dust (2017)—Full Transcript," July 26, 2017, *Scraps from the Loft*, https://scrapsfromtheloft.com/comedy/sarah-silverman-a-speck-of-dust-2017/.

**"In the nut forests of America":** Ernest Thompson Seton, *Bannertail: The Story of a Graysquirrel* (Charles Scribner's Sons, 1922), vi.

**Evidence of a seed dispersal crisis:** Sara Beatriz Mendes et al., "Evidence of a European Seed Dispersal Crisis," *Science* 386, no. 6718 (October 10, 2024): 206–11, https://doi.org/10.1126/science.ado1464.

**"Not much goes on in the mind of a squirrel":** Kate DiCamillo, *Flora and Ulysses: The Illuminated Adventures* (Candlewick Press, 2013), 10.

**great deal we are uncovering:** Nancy F. Castaldo, *Beastly Brains: Exploring How Animals Think, Talk, and Feel* (Houghton Mifflin Harcourt, 2017).

**This ratio, comparable to that of many primates:** University of Edinburgh, "Bigger brains gave squirrels the capacity to move up in the world," ScienceDaily, www.sciencedaily.com/releases/2021/04/210412114812.htm

**Squirrels use the method of "spatial chunking":** Pizza Ka Yee Chow et al., "Cognitive Performance of Wild Eastern Gray Squirrels (*Sciurus carolinensis*) in Rural and Urban, Native, and Non-Native Environments," *Frontiers in Ecology and Evolution* 9 (February 26, 2021), https://doi.org/10.3389/fevo.2021.615899.

**Michael Steele, along with researchers:** Pizza Ka Yee Chow et al., "Cognitive Performance of Wild Eastern Gray Squirrels (*Sciurus carolinensis*) in Rural and Urban, Native, and Non-Native Environments," *Frontiers in Ecology and Evolution* 9 (February 26, 2021), https://doi.org/10.3389/fevo.2021.615899.

**"No longer wabbly or vague":** Seton, *Bannertail*, 253–54.

**"I am informed by A. K. Fisher":** Ernest Thompson Seton, *Life-Histories of Northern Animals: An Account of the Mammals of Manitoba* (Charles Scribner's Sons, 1909), 320.

**"The autumn of 1851":** Winslow Cossoul Watson, *The Military and Civil History of the County of Essex, New York* (Wentworth Press, 2019), 350.

**"The opinions of people, are, however":** Robert T. Hatt, "The Red Squirrel: Its Life History and Habits, with Special Reference to the Adirondacks of New York and the

Harvard Forest," *Roosevelt Wild Life Annals* 2, no. 1 (March 1929): 12, Roosevelt Wild Life Forest Experiment Station.

**seven times as many walnuts germinate:** Robert K. Swihart and Jacob R. Goheen, "Food-Hoarding Behavior of Gray Squirrels and North American Red Squirrels in the Central Hardwoods Region: Implications for Forest Regeneration," *Canadian Journal of Zoology* 81, no. 9 (September 2003): 1636–39, https://doi.org/10.1139/Z03-143.

**"On the 24th of September in 1857":** Henry David Thoreau, "The Succession of Forest Trees," in *Faith in a Seed: The Dispersion of Seeds and Other Late Natural History Writings*, ed. Bradley Dean (Island Press, 1996), 106–7.

**"germination experiments revealed equal":** Michael A. Steele et al., "Tannins and Partial Consumption of Acorns: Implications for Dispersal of Oaks by Seed Predators," *American Midland Naturalist* 130, no. 2 (October 1993): 229–38, https://doi.org/10.2307/2426123.

**free-ranging eastern gray squirrels:** Peter D. Smallwood and Wm. David Peters, "Grey Squirrel Food Preferences: The Effects of Tannin and Fat Concentration," *Ecology* 67, no. 1 (February 1986): 168–74, https://doi.org/10.2307/1938515.

**What appears a random act:** Mikel M. Delgado and Lucia F. Jacobs, "Caching for Where and What: Evidence for a Mnemonic Strategy in a Scatter-Hoarder," *Royal Society Open Science* 4, no. 9 (September 13, 2017): 170958, https://doi.org/10.1098/rsos.170958.

**"I can find no published scientific studies":** Joe Boggs, "Tips on Oak Tips," *Buckeye Yard & Garden Online*, Ohio State University, October 13, 2021, https://bygl.osu.edu/index.php/node/1882.

**reports of oaks losing branch tips in 2018 that revealed tooth marks:** Linda Williams, "Oak Branch Tips Laying on the Ground This Fall," *Wisconsin DNR Forestry News*, September 26, 2018, https://forestrynews.blogs.govdelivery.com/2018/09/26/oak-branch-tips-laying-on-the-ground-this-fall/.

**These strong ecological relationships:** Jacob W. Dittel et al., "Reciprocal Pilfering in a Seed-Caching Rodent Community: Implications for Species Coexistence," *Behavioral Ecology and Sociobiology* 71, no. 147 (2017): 1–10, https://doi.org/10.1007/s00265-017-2375-4.

**"We would put little transmitters on the squirrels":** Jeff Gagnon, interview by the author, Arizona, April 24, 2025.

**inhabit the four different layers of the rainforest:** Konstans Wells et al., "Movement Trajectories and Habitat Partitioning of Small Mammals in Logged and Unlogged Rain Forests on Borneo," *Journal of Animal Ecology* 75, no. 5 (2006): 1212–23, https://doi.org/10.1111/j.1365-2656.2006.01144.x.

**Big Data models "are especially important":** Moritz Steiner et al., "With Super SDMs (Machine Learning, Open Access Big Data, and the Cloud), Towards More Holistic Global Squirrel Hotspots and Coldspots," *Scientific Reports* 14, no. 5204 (2024), https://doi.org/10.1038/s41598-024-55173-8.

"habitat needs and ranges are widely unknown": Steiner et al., "With Super SDMs."

A 2023 study addressed this precisely: Swapna Nelaballi, "Exploring Fruit-Frugivore Interactions and Post-Dispersal Seed Fate of Large-Seeded Plants in the Lowland Rainforests of Indonesian Borneo" (PhD diss., University of Michigan, 2023), https://hdl.handle.net/2027.42/192328.

crucial food source for hawks, foxes, coyotes: Laura M., "The Potgut Paradox: Are Uinta Ground Squirrels Pests or Eco-Warriors?," *TownLift: Park City News*, March 22, 2024, https://townlift.com/2024/03/the-potgut-paradox-are-uinta-ground-squirrels-pests-or-protectors/.

"A ferret doesn't eat anything but a prairie dog": Jessica Simmons, interview by the author, Arizona, April 24, 2025.

"the big Demon of Commerce has declared war": Ernest Thompson Seton, *Lives of Game Animals*, vol. 2 (Doubleday, Doran & Company, 1929), 573–74.

Leopards (Panthera pardus): Anirudh Nair, "Hop, Skip and Jump: The Giant Squirrels of India," *Roundglass Sustain*, January 20, 2021, updated September 24, 2023, https://roundglasssustain.com/photo-stories/giant-squirrels-of-india.

### Chapter 2

"The world is watching": Joseph Biden, State of the Union Address, The White House, March 7, 2024, https://bidenwhitehouse.archives.gov/state-of-the-union-2024/.

among the 294 squirrel species: International Union for Conservation of Nature and Natural Resources (IUCN), Red List of Threatened Species, version 2025-1, accessed May 14, 2025, https://www.iucnredlist.org/search/list?query=Squirrels&searchType=species.

"sun never sets on the Sciuridae!": Richard W. Thorington Jr. and Katie Ferrell, *Squirrels: The Animal Answer Guide* (Johns Hopkins University Press, 2006), xiv.

They air-dropped endemic pocket gophers: Jules Bernstein, "How Gophers Brought Mount St. Helens Back to Life in One Day," University of California, November 7, 2024, https://www.universityofcalifornia.edu/news/how-gophers-brought-mount-st-helens-back-life-one-day.

researchers recorded forty thousand plants: Bernstein, "How Gophers Brought Mount St. Helens Back to Life in One Day."

"You don't encounter them": Frank Carini and Colleen Cronin, "The Secret Is Out: Some Local Squirrels Can Fly," *ecoRI News*, January 5, 2024, https://ecori.org/the-secret-is-out-some-local-squirrels-can-fly/.

"Southern flying squirrels are typically": Haley O'Brien, "The Search for Endangered Flying Squirrels in the Poconos," *The Allegheny Front*, February 16, 2024, updated March 6, 2024, https://www.alleghenyfront.org/pennsylvania-poconos-endangered-northern-flying-squirrels/.

**recent study of threatened Persian squirrels:** Yiannis V. Zevgolis et al., "Ecological Implications of Deep Pruning: A Case Report on Persian Squirrel Nesting in a Centennial Olive Grove on the Island of Lesvos, Greece," *Historia Naturalis Bulgarica* 46, no. 3 (March 1, 2024): 89–97, https://doi.org/10.48027/hnb.46.034.

**"the prairie dog has been cussed":** Matt Jaffe, "Treated like a Dog!," *Arizona Highways*, accessed November 12, 2024, https://www.arizonahighways.com/article/treated-dog.

**the 2015 translocation protocol:** Holly Hicks et al., "Translocation Protocol for Gunnison's Prairie Dogs in Arizona," Nongame and Endangered Wildlife Program Technical Report 293, Arizona Game and Fish Department, https://habitatharmony.org/docs/translocation-protocol-for-gpd.pdf.

**the 2018 guide to nonlethal management:** *A Non-Lethal Management Guide for Gunnison's Prairie Dogs* (Habitat Harmony, 2018), https://habitatharmony.org/handbook.

**"They only have one litter of pups":** Emily Renn, interview by the author, April 24, 2025.

**"From deer stands, I have often watched":** Steve Gilliland, "Exploring Kansas Outdoors: Squirrels; Love 'Em or Hate 'Em," *Hays Post*, February 11, 2024, https://hayspost.com/posts/16895b94-700a-4729-ac29-8eb149aa048b.

**"I asked the chair":** Jerry Bohnen, "How a Snowball Plays a Role in Sen. Inhofe's Legacy," *Oklahoma Energy Today*, February 26, 2022, https://www.okenergytoday.com/2022/02/how-a-snowball-plays-a-role-in-sen-inhofes-legacy/.

**"Protecting wildlife is critical to fostering":** "Bravo! Governor Hochul Signs New Law to Prohibit Wildlife Killing Contests," Protect the Adirondacks!, press release, December 23, 2023, https://www.protectadks.org/bravo-governor-hochul-signs-new-law-to-prohibit-wildlife-killing-contests/.

**at least 57,000 animals:** Humane World for Animals, "Wildlife Killing Contests," accessed May 14, 2025, https://www.humaneworld.org/en/issue/wildlife-killing-contests.

**"Today is a win for every animal":** "New York Bans Wildlife Killing Contests, Becomes Tenth State to Do So," Wolf Conservation Center, December 22, 2023, https://nywolf.org/2023/12/new-york-bans-wildlife-killing-contests-becomes-tenth-state-to-do-so/.

**The new law makes it illegal:** New York Environmental Conservation Law § 11-0901 (14), https://newyork.public.law/laws/n.y._environmental_conservation_law_section_11-0901; New York State, "Governor Hochul Signs Legislation to Protect Wildlife," news release, December 22, 2023, https://www.governor.ny.gov/news/governor-hochul-signs-legislation-protect-wildlife; "Bravo! Governor Hochul Signs New Law to Prohibit Wildlife Killing Contests."

**"We support hunting that's done in a sustainable":** Alex Brown, "No More Prizes for Killing 'Nuisance' Animals Under These Hunting Contest Bans," *New Hampshire*

*Bulletin*, January 30, 2024, https://newhampshirebulletin.com/2024/01/30/no-more-prizes-for-killing-nuisance-animals-under-these-hunting-contest-bans/.

**"No person shall have in possession":** N.J. Admin. Code § 7:25-5.22, "Wild Animals; possession, killing," https://www.law.cornell.edu/regulations/new-jersey/N-J-A-C-7-25-5-22.

**"I like to go out there and just bring my gun":** Lucy Grindon, "One Last Rabbit Hunt in Norfolk Before Some Hunting Competitions Are Banned," NCPR, January 23, 2024, https://www.northcountrypublicradio.org/news/story/49121/20240123/one-last-rabbit-hunt-in-norfolk-before-some-hunting-competitions-are-banned.

**"These contests result in the wasteful":** *NCPR News*, "1/23/24: Goodbye, Rabbit and Squirrel Hunting Tournaments," NCPR, January 23, 2024, https://www.northcountrypublicradio.org/news/story/49162/20240123/1-23-24-goodbye-rabbit-and-squirrel-hunting-tournaments.

**"It is shocking that in New York":** "Bravo! Governor Hochul Signs New Law to Prohibit Wildlife Killing Contests."

**"It (the lesson from his father) built a deep respect":** Visnu Mohan, "'Not a Fan of People Who Kill for Sport'—When Brian Harman Opened Up About Hunting and the Life Lessons It Gave," *Sportskeeda*, July 24, 2023, https://www.sportskeeda.com/golf/news-not-fan-people-kill-sport-when-brian-harman-opened-hunting-life-lessons-gave.

**wildlife is a renewable resource:** Texas Parks and Wildlife Department, "Hunter Education Online Course: Chapter 9—Wildlife Conservation," https://tpwd.texas.gov/education/hunter-education/online-course/wildlife-conservation.

**Louisiana's Lafayette Advertiser warned:** Terry L. Jones, "Alligator Tales: Stories from Louisiana's Alligator-Filled Past," *Louisiana Sportsman*, September 1, 2012, https://www.louisianasportsman.com/hunting/other-hunting/alligator-tales-stories-from-louisianas-alligator-filled-past/.

**"approximately 3.5 million Louisiana alligators were killed":** Jones, "Alligator Tales."

## Chapter 3

**"Humans and nature construct one another":** Alexander Wilson, *The Culture of Nature: North American Landscape from Disney to the Exxon Valdez* (Blackwell, 1992), 13.

**"No other animal":** Diane Ackerman, "In Praise of Squirrels," *National Geographic* 188, no. 5 (November 1995).

**"front-row seat":** Ackerman, "In Praise of Squirrels."

**"I sit on a fallen trunk":** Annie Dillard, *Pilgrim at Tinker Creek* (Harper's Magazine Press, 1974).

**The legend explains how:** Christopher J. Yahnke, "Xylem Sap Moon," Squirrel-Net, April 19, 2023, https://www.squirrel-net.org/post/xylem-sap-moon.

**The tooth marks left:** Bernd Heinrich, "Maple Sugaring by Red Squirrels," *Journal of Mammalogy* 73, no. 1 (April 14, 1992): 51–54, https://doi.org/10.2307/1381865.

**"Squirrels also foresee a storm":** Pliny the Elder, in *Natural History*, Book 8, 58, *The Medieval Bestiary*, accessed 2023, https://bestiary.ca/beasts/beastsource105918.htm.

**"Nature is what we see—":** Emily Dickinson, *The Complete Poems of Emily Dickinson*, ed. Thomas H. Johnson (Back Bay Books, 1976), 668.

**"The submissions convinced me that these young poets":** John Kelly, quoted in Christina Barron, "Haiku Explored the Highs and Lows of Life as a Squirrel," *The Washington Post*, April 14, 2019, https://www.washingtonpost.com/lifestyle/kidspost/haikus-explored-the-highs-and-lows-of-life-as-a-squirrel/2019/04/12/b48354b6-5563-11e9-814f-e2f46684196e_story.html.

**"Ratatoskr is the squirrel":** Henry Adams Bellows, *The Poetic Edda: The Heroic Poems* (Loki's Publishing, 2014).

**"You little thing! You're in my way":** Madhu Chanda Das, "The Squirrel and Hanuman: Lessons from the *Ramayana*," August 14, 2024, https://www.bhagavatam-katha.com/ramayana-story-little-squirrel-who-helped-lord-rama/.

**"I want people, especially children":** Udbhavi Balakrishna, "Bengaluru Sculptor's Symbolic Squirrel Statue to Greet Passengers at Ayodhya Station," *Deccan Herald*, January 10, 2024, https://www.deccanherald.com/india/karnataka/bengaluru/city-sculptor-s-symbolic-squirrel-statue-to-greet-passengers-at-ayodhya-station-2843857.

**"The initial approach was to find someone really obscure":** Nivea Serrao, "Shannon and Dean Hale on the Unbeatable Experience of Writing *Squirrel Girl*," *Entertainment Weekly*, February 7, 2017, https://ew.com/books/2017/02/07/shannon-hale-dean-hale-unbeatable-squirrel-girl-squirrel-meets-world/.

**"Yes, I think the squirrels are great":** Maria Johansson et al., "Wildlife and Public Perceptions of Opportunities for Psychological Restoration in Local Natural Settings," *People and Nature* 6, no. 2 (April 2024): 800–817, https://doi.org/10.1002/pan3.10616.

**"The lytill squerell, full of besyness":** *Online Etymology Dictionary*, "squirrel (n.)," https://www.etymonline.com/word/squirrel.

## Chapter 4

**"We don't know a millionth":** Burton Egbert Stevenson, *Golden Book* (April 1931), according to *Stevenson's Book of Quotations*, 3rd ed. (Cassell, 1938).

**"According to the app, Eastern gray squirrels tended":** Talia Ogliore, "How Bias Shows Up in Maps Made with Citizen Science Data," Washington University in St. Louis, *The Source*, March 5, 2024, https://source.washu.edu/2024/03/how-bias-shows-up-in-maps-made-with-citizen-science-data/.

**"Squirrels are abundant in the northern part of the city":** Dr. Elizabeth Carlen (@E_Carlen), post on X (formerly Twitter), May 4, 2022, accessed April 17, 2025, https://x.com/E_Carlen/status/1521924940211302400.

**"Similar pattern to what we see in Syracuse":** John P. Vanek, PhD (@wild_ecology), post on X (formerly Twitter), May 4, 2022, accessed May 13, 2025, https://x.com/wild_ecology/status/1521948953729802241.

**often spotters don't record:** Elizabeth J. Carlen et al., "A Framework for Contextualizing Social-Ecological Biases in Contributory Science Data," *People and Nature* 6, no. 2 (March 3, 2024): 377–90, https://doi.org/10.1002/pan3.10592.

**"The International Colloquium on Squirrels is a global event":** Vincent Wildlife Trust Ireland, "8th International Colloquium on Squirrels," accessed November 23, 2024, https://www.vincentwildlife.ie/news-media/blog/8th-international-colloquium-on-squirrels.

**"cities may gain an overly positive view of citizens' attitudes":** Artti Juutinen et al., "Citizens' Attitudes Toward the Protection of Flying Squirrels in Urban Areas," *Ecology & Society* 28, no. 4 (December 26, 2023): art. 19, https://www.ecologyandsociety.org/vol28/iss4/art19/.

**invasion of Ukraine in February 2022:** International Union for Conservation of Nature and Natural Resources (IUCN), Red List of Threatened Species, 2023, https://www.iucnredlist.org; D. Radoslav et al., "Effects of Habitat Fragmentation on European Ground Squirrel (*Spermophilus citellus*) Populations in Central and Eastern Europe," *Biodiversity and Conservation* 31, no. 6 (2022): 1345–60, https://www.researchgate.net/publication/388032292_Spermophilus_citellus_European_Ground_Squirrel_The_IUCN_Red_List_of_Threatened_Species_2024; M. Shkvyria and L. Shevchenko, "War and Wildlife: The Impact of Armed Conflict on Terrestrial Mammal Habitats in Ukraine," *Ukrainian Journal of Ecology* 12, no. 4 (2022): 23–30, https://www.researchgate.net/publication/390066320_The_impact_of_armed_conflict_on_biodiversity#full-text

**"Is it unsurprising that rodents—the dominant mammal residents of steppes":** Mikhail Rusin, "Threats of Russian Invasion for Protected Small Mammals in Ukraine," Ukraine War Environmental Consequences Work Group, February 8, 2023, https://uwecworkgroup.info/threats-of-russian-invasion-for-protected-small-mammals-in-ukraine/.

**"We have more questions than answers about their social":** Donna J. Miller, "The Secret Lives of Squirrels Being Researched at Baldwin Wallace University: Animals in the News," Cleveland.com, August 7, 2012, https://www.cleveland.com/metro/2012/08/_animals_in_the_news_19.html.

**"One of the things we are trying to understand":** Samantha Grenrock, "Why Should You Love Squirrels? Here Are Six Reasons," *University of Florida News*, January 17, 2018, https://archive.news.ufl.edu/articles/2018/01/why-should-you-love-squirrels-here-are-six-reasons.html.

**"The species has quite a varied repertoire":** Laurie D. Morrissey, "Squirrel Talk: What Does That Noise Mean?," *Northern Woodlands*, November 28, 2022, https://northernwoodlands.org/outside_story/article/squirrel-talk.

**"These buggers are cranking out reproduction":** Ned Rozell, "The Secret Life of Red Squirrels," University of Alaska Fairbanks, *UAF News and Information*, November 27, 2024, https://www.uaf.edu/news/the-secret-life-of-red-squirrels.php.

**the neon pink that black light picks up in squirrel fur:** Bryan Hughes et al., "Using Mass Spectrometry to Investigate Fluorescent Compounds in Squirrel Fur," *PLoS One* 17, no. 2 (2022): e0257156, https://doi.org/10.1371/journal.pone.0257156.

**"sentinel for pathologies that have environmental causes":** Caterina Raso et al., "Ectopic Pregnancy and T-Cell Lymphoma in a Eurasian Red Squirrel (*Sciurus vulgaris*): Possible Comorbidity and a Comparative Pathology Perspective," *Animals* (Basel) 14, no. 5 (February 27, 2024): 731, https://doi.org/10.3390/ani14050731.

## Chapter 5

**"He ate various animals":** Delaware Tribe of Indians, "Stories—Këkhìtkil Hùnt Na Xanikw: Squirrels Were Said to Be Huge," *Lenape Talking Dictionary*, accessed December 3, 2024, https://www.talk-lenape.org/stories?id=27.

**cultural education director for the Delaware Tribe of Indians:** Jeremy Johnson, information provided to visitors of the Hawk Mountain Sanctuary, April 4, 2024.

**"greene country towne":** Inga Saffron, "Green Country Town," *Encyclopedia of Greater Philadelphia*, accessed September 1, 2022, https://philadelphiaencyclopedia.org/themes/green-country-town/.

**"being the largest and smelling like a Fox":** John Brickell, *The Natural History of North Carolina: With an Account of the Trade, Manners, and Customs of the Christian and Indian Inhabitants. Illustrated with a Map, and Copper-Plates, etc.* (Charles Corbett, 1743), 127.

**"much the same Virtues":** Brickell, *Natural History of North Carolina*, 129.

**"A correspondent from Johnston informs":** *The Gazette of the United-States*, June 18, 1791, page 59, image 3, https://chroniclingamerica.loc.gov/lccn/sn83030483/1791-06-18/ed-1/seq-3/. **"This species . . . was once so excessively multiplied":** John Davidson Godman, *American Natural History*, vol. 2 (Carey, Lea & Carey, 1828), 131, https://archive.org/details/americannaturalh2182godm/.

**"This species is remarkable among all our squirrels":** Godman, *American Natural History*, 132.

**"ground squirrel, or little striped squirrel":** William Bartram, *The Travels of William Bartram* (Macy-Masius, 1928), 233, https://books.google.com/books/about/The_Travels_of_William_Bartram.html?id=-u1HAAAAIAAJ.

**"This animal is so well known":** Samuel N. Rhoads, *The Mammals of Pennsylvania and New Jersey: A Biographic, Historic and Descriptive Account* (privately published, Philadelphia, 1903), 52, https://archive.org/details/mammalsofpennsyloorhoaiala.

**"Squirrels might seem too small and commonplace"**: Dan Flores, *Wild New World: The Epic Story of Animals and People in America* (W. W. Norton, 2022).

**"A correspondent calls our attention"**: *Philadelphia Ledger*, May 11, 1847, accessed April 2024 at the Historical Society of Pennsylvania.

**"With fountains glancing"**: *Philadelphia Ledger*, May 11, 1847.

**"probably our best-known"**: Vernon Bailey, "Animals Worth Knowing Around the Capitol," 1934, unpublished manuscript, P. 1, Folder 5: "Animals Worth Knowing Around the National Capitol. Text of a Radio Talk over WRC, July 12, 1934," Box 7, Record Unit 7267, Vernon Orlando Bailey Papers 1889–1941 and undated (Smithsonian Institution Archives, Washington, DC).

**"A wide variety of institutions and individuals"**: Etienne Benson, "The Urbanization of the Eastern Gray Squirrel in the United States," *Journal of American History* 100, no. 3 (December 2013): 691–710, https://doi.org/10.1093/jahist/jat353.

**"cure boys of their tendency toward cruelty"**: Benson, "Urbanization of the Eastern Gray Squirrel in the United States."

**"elegant creature"**: John Burroughs, *Squirrels and Other Fur-Bearers* (Houghton Mifflin, 1875).

**"fully a hundred fine fat squirrels"**: Sadie Stein, "Alien Squirrel," *New York Magazine*, February 3, 2014, https://nymag.com/news/features/squirrels-2014-2/.

**"A round dozen, by actual count"**: "Red Breasts And Bushy Tails: A Glance at the Robins and the Gray Squirrels in Central Park," *The New York Sun*, June 4, 1888, 2. https://chroniclingamerica.loc.gov/lccn/sn83030272/1888-06-04/ed-1/seq-2/#date1=18 88&index=0&rows=20&words=Central+Park+squirrel&searchType=basic&sequence =0&state=New+York&date2=1888&proxtext=Central+Park+squirrel&y=0&x=15&da teFilterType=yearRange&page=1

**"Endless Source of Delight to Visitors"**: "Central Park Squirrels: An Endless Source of Delight to Visitors, Young and Old," *The New York Sun*, October 21, 1900, image 17, https://chroniclingamerica.loc.gov/lccn/sn83030272/1900-10-21/ed-1/seq-17/.

**"discovered that the children's summer friends"**: *New-York Daily Tribune*, "Willcox Squirrels' Friend: The New Commissioner Provides Peanuts for Hungry Park Dwellers," January 11, 1902, https://chroniclingamerica.loc.gov/lccn/sn83030214/1902-01-11 /ed-1/seq-10.

**"none other than a superb rustic apartment house"**: Eugene Kinkead, "The Squirrels of Central Park," *The New Yorker*, September 9, 1974 (original report dated 1907), 101, https://www.newyorker.com/magazine/1974/09/09/the-squirrels-of-central-park.

**"In one respect squirrels do not differ from men"**: "Squirrels and Men," *The Evening World's Daily Magazine*, December 18, 1906, https://www.nyshistoricnewspapers.org/.

**"Mr. Wirth has confidence they"**: Susan Du, "Did a Famed Parks Leader Import Gray Squirrels to Minneapolis—and Have the Red Ones Killed?" *Minnesota Star*

*Tribune*, January 20, 2023, https://www2.startribune.com/gray-red-squirrels-theo dore-wirth-loring-park-birds-minneapolis/600245052/.

**"the most famous squirrel ever to come from Washington":** John Kelly, "Tommy Tucker, Washington's Most Famous Squirrel," *The Washington Post*, April 8, 2012, https://www.washingtonpost.com/local/tommy-tucker-washingtons-most-famous -squirrel/2012/04/08/gIQAddnZ4S_story.html.

**"salmon of the interior West":** Mike Koshmrl, "Chislers Back on Top After Wintering Underground," *Jackson Hole News & Guide*, April 25, 2018, https://www.jhnews andguide.com/valley/feature/chislers-back-on-top-after-wintering-underground/ar ticle_18315559-15e6-5394-9ba9-8a55fd6c6e7b.html.

**"unforgiving" and "absolutely ferocious":** "Rock Squirrels Considered the Most Dangerous Animal at the Grand Canyon," Grand Canyon Helicopter Tours, accessed April 17, 2025, https://grandcanyonhelicoptertour.net/rock-squirrels-considered-the -most-dangerous-animal-at-the-national-grand-canyon/.

### Chapter 6

**"A Walmart greeter is an employee":** Sarah Nassauer, "Welcome Back, Wal-Mart Greeters," *The Wall Street Journal*, June 18, 2015, https://www.wsj.com/articles/wal -mart-ushers-greeters-back-to-the-front-1434651944.

**"When I first started the project":** Noah Perlut, interview by the author, October 9, 2024.

**"Nobody knew anything about them":** Perlut, interview.

**"I wanted the students to walk":** Perlut, interview.

**"I think a big thing about this class":** Autumn Linden, interview by the author, October 9, 2024.

**"One of the fun things about this class":** Rick Hamilton, interview by the author, October 9, 2024.

**"an absolute sweetheart":** Hamilton, interview.

**"You see them everywhere":** Hamilton, interview.

### Chapter 7

**"When Timmy and Goody Tiptoes":** Beatrix Potter, "The Tale of Timmy Tiptoes," in *The Complete Tales of Beatrix Potter* (F. Warne & Co., 1989), 238.

**"flattening diversity":** Peter Coates, *Squirrel Nation: Reds, Greys and the Meaning of Home* (Reaktion Books, 2023).

**"The Squirrels came safe and well":** "From Benjamin Franklin to Deborah Franklin, 28 January 1772," National Archives and Records Administration, *Founders Online*, accessed January 1, 2025, https://founders.archives.gov/documents/Franklin/01-19-02 -0028.

Source Notes

**"Alas! Poor Mungo!":** "From Benjamin Franklin to Georgiana Shipley, 26 September 1772," National Archives and Records Administration, *Founders Online*, accessed January 1, 2025, https://founders.archives.gov/documents/Franklin/01-19-02-0202.

**"its favorite thing was to jump":** "Louisiana Squirrel," hand-colored drawing, 1727, Library of Congress, accessed January 1, 2025, https://www.loc.gov/item/2021668135/.

**this "midden-gifting" behavior:** J. E. Lane et al., "Post-Weaning Parental Care Increases Fitness but Is Not Heritable in North American Red Squirrels," *Journal of Evolutionary Biology* 28, no. 6 (June 2015): 1203–12, https://doi.org/10.1111/jeb.12633.

**"I knew quite a bit about the war":** Jane Yolen, "Trash Mountain," accessed May 15, 2025, https://www.janeyolen.com/trash-mountain/.

**"larger, faster, and more aggressive":** Jane Yolen, *Trash Mountain* (Lerner, 2015).

**"This you should know":** Yolen, *Trash Mountain*.

**Roughly seventy different definitions of intelligence:** Dan Nosowitz, "Is a Squirrel Smarter than a Fifth-Grader?," *Atlas Obscura*, August 13, 2015, https://www.atlasobscura.com/articles/is-a-squirrel-smarter-than-a-fifthgrader.

**"When scatter hoarding a food item":** Michael A. Steele et al., "Cache Protection Strategies of a Scatter-Hoarding Rodent: Do Tree Squirrels Engage in Behavioural Deception?," *Animal Behaviour* 75, no. 2 (February 2008): 705–14, https://doi.org/10.1016/j.anbehav.2007.07.026.

**The squirrel will dig a hole:** Nosowitz, "Is a Squirrel Smarter than a Fifth-Grader?"

**squirrel deception in cache protection strategies:** Jaeeun Shim and Ronald C. Arkin, "Robot Deception and Squirrel Behavior: A Case Study in Bio-Inspired Robotics," Georgia Institute of Technology, accessed May 15, 2025, https://sites.cc.gatech.edu/ai/robot-lab/online-publications/squirrel_preprint_shim14.pdf; Jaclyn R. Aliperti et al., "Bridging Animal Personality with Space Use and Resource Use in a Free-Ranging Population of an Asocial Ground Squirrel," *Animal Behaviour* 180 (October 2021): 291–306, https://doi.org/10.1016/j.anbehav.2021.07.019.

**"Her chariot is an empty hazel-nut":** William Shakespeare, "Romeo and Juliet," Act 1, Scene 4, Folger Shakespeare Library, accessed January 1, 2025, https://www.folger.edu/explore/shakespeares-works/romeo-and-juliet/read/1/4/.

**"And to this day, if you meet Nutkin":** Beatrix Potter, *The Tale of Squirrel Nutkin* (F. Warne, 1987).

**"Grey squirrels are not as crazy invaders as we think":** Hayley Dunning, "Don't Blame Grey Squirrels: Their British Invasion Had Much More to Do with Us," Imperial College London, *Imperial News*, January 26, 2016, https://www.imperial.ac.uk/news/170448/dont-blame-grey-squirrels-their-british/.

**"Red squirrels are now an endangered species in the UK":** University of Surrey, "Scientists Are Unravelling the Secrets of Red and Grey Squirrel Competition," press release, February 15, 2024, https://www.surrey.ac.uk/news/scientists-are-unravelling-secrets-red-and-grey-squirrel-competition.

**"formed by communities of micro-organisms":** Lucy Hall et al., "Significant Differences in the Caecal Bacterial Microbiota of Red and Grey Squirrels in Britain," *Journal of Medical Microbiology* 73, no. 2 (February 14, 2022): 001793, https://doi.org/10.1099/jmm.0.001793.

**could cost at least £1.1 billion:** Royal Forestry Society, Forestry Commission, Natural Resources Wales, National Forest Company, and Woodland Trust, "An Analysis of the Cost of Grey Squirrel Damage to Woodland," January 2021, https://rfs.org.uk/wp-content/uploads/2021/03/analysis-of-the-cost-of-grey-squirrel-damage-to-woodland-publication-copy-180121.pdf.

**2021 fact sheet:** UK Squirrel Accord and Animal and Plant Health Agency, "Grey Squirrel Fertility Control Research: Frequently Asked Questions—June 2021," https://squirrelaccord.uk/wp-content/uploads/2024/08/UKSA_fertility_control_research_FAQs_June_2021_-_research_-_UK_Squirrel_Accord.pdf.

**"also has implications for human health":** Hall et al., "Significant Differences in the Caecal Bacterial Microbiota."

**"When the grays show up, it puts the reds":** Marlena Spieler and *The New York Times*, "Squirrel Finds New Popularity Among British Diners," *The Mercury News*, February 3, 2009, https://www.mercurynews.com/2009/02/03/squirrel-finds-new-popularity-among-british-diners/.

**"These days, however, in farmers' markets":** Marlena Spieler, "Saving a Squirrel by Eating One," *The New York Times*, January 6, 2009, https://www.nytimes.com/2009/01/07/dining/07squirrel.html.

**"their lovely flavor":** Spieler, "Saving a Squirrel by Eating One."

**"While some have difficulty":** Spieler, "Saving a Squirrel by Eating One."

**"How could you?":** Food Urchin, "Potted Squirrel with Sourdough," *Great British Chefs*, August 8, 2024, https://www.greatbritishchefs.com/recipes/potted-squirrel-recipe.

**"As you all know so well, these charming":** Caroline Hallemann, "Prince Charles Once Again Proves He's the Red Squirrel's Number One Fan," *Town & Country*, January 21, 2021, https://www.townandcountrymag.com/society/tradition/a35280072/prince-charles-red-squirrels-message/.

**"finer textured and has a more subtle flavour":** Wild Meat Company, "Squirrel," accessed January 1, 2025, https://www.wildmeat.co.uk/products/squirrel.

**"We choose food sources":** Douglas McMaster, "'Zero Waste Is Just a System with No Loose Ends,'" DouglasMcMaster.com, accessed December 13, 2024, https://www.douglasmcmaster.com/silolondon.

**"creatively popularize species":** Daniel Matthews and Sylvain Peuchmaurd, "Invasive Species on the Menu at London Restaurant," *Taipei Times*, September 21, 2023, https://www.taipeitimes.com/News/feat/archives/2023/09/21/2003806548.

**article published by The Royal Society:** Emma Sheehy et al., "The Enemy of My Enemy Is My Friend: Native Pine Marten Recovery Reverses the Decline of the Red Squirrel by Suppressing Grey Squirrel Populations," *Proceedings of the Royal Society B: Biological Sciences*, March 7, 2018, https://doi.org/10.1098/rspb.2017.2603.

**"They've managed to co-exist":** Victoria Gill, "Red Squirrel Numbers Boosted by Predator," *BBC News*, March 6, 2018, https://www.bbc.com/news/science-environment-43308588.

**"Where pine marten activity is high":** Gill, "Red Squirrel Numbers Boosted by Predator."

**"Pine marten predation thus reverses":** Sheehy et al., "The Enemy of My Enemy Is My Friend."

**"Somebody told me at the beginning":** Amy Houghton, "Rare (and Very Cute) Baby Red Squirrels Have Been Born in Yorkshire." *Time Out United Kingdom*, August 11, 2023, https://www.timeout.com/uk/news/rare-and-very-cute-baby-red-squirrels-have-been-born-in-yorkshire-081123.

**"offer a humane, efficient, species-specific":** Nicky R. Faber et al., "Novel Combination of CRISPR-Based Gene Drives Eliminates Resistance and Localises Spread," *Scientific Reports* 11, art. 3719 (2021), https://doi.org/10.1038/s41598-021-83239-4.

**"We don't want to demonise them":** Catherine Mackinlay, "*Cumbrian Red* Filmmaker Talks 'Scamps' Ahead of South Cumbria Showings," *The Mail*, February 11, 2024, https://www.nwemail.co.uk/news/24104080.cumbrian-red-filmmaker-talks-scamps-ahead-south-cumbria-showings/.

**"I'm honestly gutted":** Terry Abraham (@terrybnd), post on X (formerly Twitter), February 2, 2024, accessed January 1, 2025, https://x.com/terrybnd/status/1753783740663226581.

**"Today it was revealed that some restaurants":** Annieli, "Ukraine Invasion Day 379: Russia Might Capture a City They Have Completely Demolished" *Daily Kos*, March 8, 2023, https://www.dailykos.com/stories/2023/3/8/2156958/-Ukraine-Invasion-Day-379-Russia-might-capture-a-city-they-have-completely-demolished.

**"Russian propagandist Skabeyeva says":** Anton Gerashchenko (@Gerashchenko_en), post on X, formerly Twitter, March 7, 2023, accessed April 2, 2025, https://x.com/Gerashchenko_en/status/1633136675567153153.

**"Indeed, the issue is the very presence of grey squirrels":** Barney Davis, "MP Labels Grey Squirrels 'Hamas of the Squirrel World,'" *The Independent* (UK), November 29, 2023, https://www.independent.co.uk/news/uk/home-news/grey-squirrels-hamas-dup-mp-b2455474.html.

**even a "cheeky" gray squirrel:** Andrew Forgrave, "Cheeky Squirrel Spotted on Snowdonia Peak Looking for Easy Pickings from Walkers," *North Wales Live*, August 3, 2023, https://www.dailypost.co.uk/news/north-wales-news/cheeky-squirrel-spotted-snowdonia-peak-27449773.

**the program can detect different species:** Genysys Engine, "Squirrel Agent: Project Red Haven," accessed November 25, 2024, https://genysysengine.tech/squirrel-feeding-agent.

**"Direct risk to nontarget fish and wildlife":** US Department of Agriculture, "The Use of GonaCon in Wildlife Damage Management," chap. 11 in *Human Health and Ecological Risk Assessment for the Use of Wildlife Damage Management Methods by USDA APHIS Wildlife Services*, August 2017, Peer Reviewed Final July 2022, https://www.aphis.usda.gov/sites/default/files/11-gonacon.pdf.

## Chapter 8

**"For us to go to Italy":** D. H. Lawrence, "To Nuoro," chap. 6 in *Sea and Sardinia* (Thomas Seltzer, 1921), http://www.online-literature.com/dh_lawrence/sea-and-sardinia/6/.

**"How beautiful is sunset":** Shaun MacLoughlin, "English Wordplay—Listen and Enjoy: Percy Bysshe Shelley, Journey to Italy and *The Cenci*," *English Wordplay*, accessed November 9, 2024, http://www.englishwordplay.com/shelley6.html.

**"Everywhere in Torino":** Sandro Bertolino, interview by the author, Torino, Italy, October 16, 2024.

**"some exotic species":** Parchi Reali Torino, "Fauna," December 17, 2024, https://parchireali.it/parco-naturale-di-stupinigi/fauna/.

**"best loved by children":** Genoa City Tours, "Nervi Parks—Serra Gropallo—Nervi," accessed January 2, 2025, https://www.genoa.in/genoa/what-to-see/details/parchi-di-nervi-serra-gropallo.

**"Squirrels proved to be successful invaders":** Sandro Bertolino, "Animal Trade and Non-Indigenous Species Introduction: The World-Wide Spread of Squirrels," *Diversity and Distributions* 15, no. 4 (June 10, 2009): 701–8, https://doi.org/10.1111/j.1472-4642.2009.00574.x.

**"Monza was an oasis":** Francesca Sironi, online interview by the author, June 13, 2024.

**"We see them":** Adam Rose, interview by the author, October 19, 2024.

**"little or no information is available":** Anna Lisa Signorile and Julian Evans, "Damage Caused by the American Grey Squirrel (*Sciurus carolinensis*) to Agricultural Crops, Poplar Plantations and Semi-Natural Woodland in Piedmont, Italy," *Forestry: An International Journal of Forest Research* 80, no. 1 (January 2007): 89–98, https://doi.org/10.1093/forestry/cpl044.

**"revealed precise details of illegal squirrel trade":** A. L. Signorile et al., "Using DNA Profiling to Investigate Human-Mediated Translocations of an Invasive Species," *Biological Conservation* 195 (March 2016): 97–105, https://doi.org/10.1016/j.biocon.2015.12.026.

**John Platt's explosive 2012 article:** John Platt, "Italy Faces Invasion of American Killer Squirrels," *Scientific American*, October 5, 2012, https://www.scientificamerican.com/blog/extinction-countdown/italy-faces-invasion-of-american-killer-squirrels/.

**"some scientists estimate that red squirrels":** John Platt, "Italy Faces Invasion of American Killer Squirrels," *Scientific American*, October 5, 2012, https://www.scientificamerican.com/blog/extinction-countdown/italy-faces-invasion-of-american-killer-squirrels/.

**noted the spread and attempted eradication:** Sandro Bertolini and Piero Genovesi, "Spread and Attempted Eradication of the Grey Squirrel (*Sciurus carolinensis*) in Italy, and Consequences for the Red Squirrel (*Sciurus vulgaris*) in Eurasia," *Biological Conservation* 109, no. 3 (March 2003): 351–58, https://doi.org/10.1016/S0006-3207(02)00161-1.

**When Pallas's squirrels were removed:** M. V. Mazzamuto et al., "Space Invaders: Effects of Invasive Alien Pallas's Squirrel on Home Range and Body Mass of Native Red Squirrel," *Biological Invasions* 19 (2017): 1863–77, https://doi.org/10.1007/s10530-017-1396-2.

**"Red squirrels occurred at much lower densities":** Maria Vittoria Mazzamuto et al., "Interspecific Competition Between Alien Pallas's Squirrels and Eurasian Red Squirrels Reduces Density of the Native Species," *Biological Invasions* 19, no. 2 (November 3, 2016): 723–35, https://doi.org/10.1007/s10530-016-1310-3.

**"We have dormice in Sicily":** Maria Mazzamuto, interview by the author via Zoom, September 18, 2024.

**"It was interesting to approach":** Mazzamuto, interview.

**"The red is pushed":** Mazzamuto, interview.

**"avoid the transnational move":** Mazzamuto, interview.

**"That connection that the local people had":** Mazzamuto, interview.

**"They become these symbols of globalization":** Etienne Benson, interview by the author, April 3, 2024.

**"flattening diversity":** Etienne Benson, interview by the author, April 3, 2024, Philadelphia.

**Venezuelan migrant Yeison:** Valerie Gonzalez, "A Venezuelan Man and His Pet Squirrel Made It to the US Border. Now He's Preparing to Say Goodbye," *AP News*, September 22, 2023, https://apnews.com/article/immigration-border-squirrel-texas-6a1dbf8244837c87f41224e09a9dd613.

## Chapter 9

**"This food ain't fitten to eat":** Marjorie Kinnan Rawlings, "Cracker Chidlings," EnglishLiterature.Net, July 15, 2021, https://englishliterature.net/marjorie-kinnan-rawlings/cracker-chidlings.

## Source Notes

"**The true history of it is about as follows**": "The Brunswick Stew: The Originator of It and How the Dish Was First Made," *The National Republican*, September 16, 1886, https://chroniclingamerica.loc.gov/lccn/sn86053573/1886-09-16/ed-1/seq-3/.

"**When I was in college**": Jonathan Stein, "Huckabee: 'We Would Fry Squirrels in a Popcorn Popper,'" *Mother Jones*, January 17, 2008, https://www.motherjones.com/politics/2008/01/huckabee-we-would-fry-squirrels-popcorn-popper/; *New York Daily News*, "Mike Huckabee Plays Up Charm in S.C.," January 11, 2019, https://www.nydailynews.com/2008/01/17/mike-huckabee-plays-up-charm-in-sc/.

"**I can shoot quite well**": Will Lawrence, "She Can Skin and Cook a Squirrel, but Jennifer Lawrence Isn't Sure If She's Ready for Super-Stardom, She Tells Will Lawrence," *Irish Independent*, March 23, 2012, https://www.independent.ie/news/she-can-skin-and-cook-a-squirrel-but-jennifer-lawrence-isnt-sure-if-shes-ready-for-super-stardom-she-tells-will-lawrence/26835448.html.

"**three or four good sized squirrels**": F. L. Gillette and Hugo Ziemann, *The White House Cook Book: Cooking, Toilet and Household Recipes, Menus, Dinner-Giving, Table Etiquette, Care of the Sick, Health Suggestions, Facts Worth Knowing, etc. from the White House of the Gilded Age* (Smithmark Publishers, 1995).

"**Mr. Cleveland brought with him**": "Mr. Cleveland Bags a Few Squirrels," *The New York Sun*, reprinted from the *Cincinnati Enquirer*, November 18, 1888, page 6, image 6, https://chroniclingamerica.loc.gov/.

"**Few things give me the feeling**": Wade Robertson, "The Ups and Downs of Squirrel Hunting," *Olean Times Herald*, September 23, 2023, updated October 27, 2024, https://www.oleantimesherald.com/sports/the-ups-and-downs-of-squirrel-hunting/article_25b870a8-5997-11ee-9e38-b7e0211df2c1.html.

"**There is absolutely nothing wrong with squirrel meat**": Steve Gilliland, "Exploring Kansas Outdoors: Squirrels; Love 'Em or Hate 'Em." *Hays Post*, February 11, 2024, https://hayspost.com/posts/16895b94-700a-4729-ac29-8eb149aa048b.

"**One of the proudest moments of my young life**": Terry L. Jones, "A Fall Tradition," *Country Roads Magazine*, September 23, 2019, https://countryroadsmagazine.com/outdoors/Squirrel-season-memories-pastimes/.

"**I grew up in Louisiana**": Laura Shunk, "Elise Wiggins on Italian Butchery, Squirrel Farming and Cattivella," *Westword*, May 17, 2017, https://www.westword.com/restaurants/elise-wiggins-on-squirrel-farming-italian-butchery-and-cattivella-9047527.

"**more protein and nutrients per ounce**": Elise Wiggins, telephone interview by the author, July 29, 2024.

"**Up here, people say**": Shunk, "Elise Wiggins."

"**I realize it's much easier to pick up a chicken**": Jason Salzman, "'Enough Is Enough': Man Eats 'Nuisance' Squirrels from His Denver Yard," *Colorado Times Recorder*, December 23, 2021, https://coloradotimesrecorder.com/2021/12/enough-is-enough-man-eats-nuisance-squirrels-from-his-denver-yard/41949/.

## Chapter 10

**"Living off the grid":** Andrew Murfett, "Q&A with Ron Perlman," *The Sydney Morning Herald*, September 16, 2012, https://www.smh.com.au/entertainment/tv-and-radio/qa-with-ron-perlman-20120914-25vpb.html.

**"What shall I do?":** Mary Oliver, *Thirst* (Beacon Press, 2006), p. 13.

**"to protect all species":** "Rosalie Barrow Edge—Feminist, Naturalist and Conservationist," Audubon Center for Birds of Prey, April 8, 2021, https://cbop.audubon.org/news/rosalie-barrow-edge-feminist-naturalist-and-conservationist.

**"People with bird feeders bring up their frustrations":** Kate Morgan, "How (and Why) to Peacefully Coexist with Squirrels," *The Washington Post*, February 21, 2024, https://www.washingtonpost.com/home/2024/02/22/living-with-squirrels-nuisance/.

**"We have a western screech-owl box":** Terry Rich, "Those Darn Squirrels!," *Idaho Press*, January 24, 2024, https://www.idahopress.com/outdoors/those-darn-squirrels/article_abe830f8-b9a2-11ee-8c54-1b9a65141e8d.html.

**bird feeder market in the United States:** Sneha Mali, "Global Outdoor Bird Feeder Market Report 2025," *Cognitive Market Research*, March 2025, https://www.cognitivemarketresearch.com/outdoor-bird-feeder-market-report.

**"When I learned, 'You are what you eat'":** Bill Carver, in Squirrel Haters of America, Facebook.

**"Was it the gray squirrels? The red squirrels?":** Diane Ackerman, *Cultivating Delight: A Natural History of My Garden* (G. K. Hall, 2002), 63.

**"See, we have a lot of squirrels":** Author's conversation with a family friend.

**"I know this is the way of nature, kill and be killed":** Ackerman, *Cultivating Delight*, 63.

**"Both red and gray squirrels will eat birds' eggs and nestlings":** Ackerman, *Cultivating Delight*, 64.

**"The reason they are chewing on things":** Samantha Grenrock, "Why Should You Love Squirrels? Here Are Six Reasons," University of Florida News Archive, January 17, 2018, https://news.ufl.edu/articles/2018/01/why-should-you-love-squirrels-here-are-six-reasons.html.

**"A squirrel trapped in a chimney":** "Squirrel Nuisance Problems," Connecticut Department of Energy & Environmental Protection, accessed January 3, 2025, https://portal.ct.gov/deep/wildlife/nuisance-wildlife/problems-with-squirrels.

**"Don't be passive to the usurpation":** Rob Speed, "Satire: Bryan College Animals Will Overthrow Campus," *The Triangle*, Bryan College, February 23, 2024, https://www.bryantriangle.com/satire-bryan-college-animals-will-overthrow-campus/.

**"Among the mammals ground squirrels":** Rachel Carson, "Needless Havoc," chap. 7 in *Silent Spring* (Houghton Mifflin, 1962), 120.

**"'exhibited a characteristic attitude in death'":** Carson, "Needless Havoc," 127–28.

**"widely found in liver tissues of birds":** Alexander Badry et al., "Linking Landscape Composition and Biological Factors with Exposure Levels of Rodenticides and Agrochemicals in Avian Apex Predators from Germany," *Environmental Research* 193 (February 2021): 110602, https://doi.org/10.1016/j.envres.2020.110602.

**"We found rodenticide residues in liver tissues":** Badry et al., "Linking Landscape Composition and Biological Factors."

**rat poisons were detected in 84 percent of the city's dead birds:** NYC Bird Alliance, "Rodenticides," accessed April 18, 2025, https://www.nycbirdalliance.org/our-work/conservation/urban-raptors/rodenticides.

### Chapter 11

**"The behavior of squirrels during September":** Vagn Flyger, "The 1968 Squirrel 'Migration' in the Eastern United States," unpublished paper, accessed April 21, 2024, http://www.myoutbox.net/flyger.htm.

**An estimated 41 million squirrels:** Krista Conrad, "Animals That Are the Biggest Road-Kill Victims in America," *WorldAtlas*, July 4, 2020, https://www.worldatlas.com/articles/animals-that-are-the-biggest-road-kill-victims-in-america.html.

**"There were squirrel everywhere":** Hunting forum member big bore bob, comment #17 in "The Great Squirrel Migration of 1968," North Carolina Hunting and Fishing Forums, September 1, 2018, https://www.nchuntandfish.com/forums/index.php?threads%2Fthe-great-squirrel-migration-of-1968.38671%2F.

**"The behavior of squirrels during September":** Flyger, "The 1968 Squirrel 'Migration.'"

**"poorly understood and should be investigated":** Flyger, "The 1968 Squirrel 'Migration.'"

**"recent cruise on the Hudson":** Daniel Smiley, "Ecosystem Sketch no. 10: Gray Squirrels on the Move," *The Chirp* 15, no. 11, November 1968, https://www.mohonkpreserve.org/wp-content/uploads/2021/08/16-Gray-Squirrels.pdf.

**previous gray squirrel population peak":** Smiley, *The Chirp*.

**"Observed a number of squirrels swimming":** Joseph A. Mussulman, "Eastern Gray Squirrel," *Discover Lewis & Clark*, February 20, 2022, https://lewis-clark.org/sciences/mammals/eastern-gray-squirrel/.

**"believe there's any good scientific record":** "Massive Squirrel Migrations Recorded in North America," *Farm Progress*, July 21, 2006, https://www.farmprogress.com/commentary/massive-squirrel-migrations-recorded-in-north-america.

**"It didn't happen all at once, but squirrels":** Steven N. Veigel, "Why Millions of Gray Squirrels Crossed America in Swarms," *AtCharlie Chronicles*, January 13, 2023, https://www.atcharlie.com/atcharlie-chronicles/why-millions-of-gray-squirrels-crossed-america-in-swarms/.

**Researchers with the Northeast Climate Adaptation Science Center:** Michael T. Hallworth et al., "Boom and Bust: The Effects of Masting on Seed Predator Range Dynamics and Trophic Cascades," *Diversity and Distributions* 30, no. 8 (August 2024): e13861, https://doi.org/10.1111/ddi.13861; Toni L. Morelli et al., "Does Habitat or Climate Change Drive Species Range Shifts?," *Ecography* (February 17, 2025): e07560, https://doi.org/10.1111/ecog.07560.

**"We've seen climate change pushing":** Etienne Benson, interview by the author.

**"Whether species and their populations":** Catherine Finn et al., "More Losers Than Winners: Investigating Anthropocene Defaunation Through the Diversity of Population Trends," *Biological Reviews* 98, no. 5 (May 15, 2023): 1732–48, https://doi.org/10.1111/brv.12974.

**the United Nations estimates:** United Nations Environment Programme, "Nature's Dangerous Decline: 'Unprecedented' Species Extinction Rates 'Accelerating,'" press release, May 6, 2019, https://www.unep.org/news-and-stories/press-release/natures-dangerous-decline-unprecedented-species-extinction-rates.

**Its isolated population is suffering:** Washington Department of Fish and Wildlife, "Western Gray Squirrel (*Sciurus griseus*)," accessed May 4, 2025, https://www.wdfw.wa.gov/species-habitats/species/sciurus-griseus#living.

**"Their populations are going down in some regions":** Jennifer Gaeng, "Squirrel Population: How Many Are There in the World?," *A-Z Animals* (blog), updated February 27, 2025, https://a-z-animals.com/blog/squirrel-population-by-state/.

**"several dozen salmon and steelhead fish populations":** Associated Press, "Most US Endangered Species Money Goes to Handful of Species," *Voice of America*, December 30, 2023, https://www.voanews.com/a/most-us-endangered-species-money-goes-to-handful-of-species-/7418303.html.

**"The issue is not where the money is spent":** Associated Press, "Most US Endangered Species Money Goes to a Handful of Species."

**"We are in the midst of an unprecedented biodiversity crisis":** Representative Debbie Dingell, "Dingell's Recovering America's Wildlife Act Advances to House Floor," press release, January 19, 2022, https://debbiedingell.house.gov/news/documentsingle.aspx?DocumentID=3349.

**The bill passed the House (231–190) in June 2022:** Recovering America's Wildlife Act of 2022, H.R. 2773, 117th Congress (2021–22) (enacted), https://www.congress.gov/bill/117th-congress/house-bill/2773.

**"RAWA represents a strong commitment":** Dingell, "Dingell's Recovering America's Wildlife Act Advances to House Floor."

**"There are at least 12,000 different species":** Alex Brown, "A Bipartisan Push Could Change State Wildlife Protection," *Stateline*, June 6, 2023, https://stateline.org/2021/03/05/a-bipartisan-push-could-change-state-wildlife-protection/.

**Trump implemented 145 actions:** Oliver Milman, "Trump Has Launched More Attacks on the Environment in 100 Days than His Entire First Term," *The Guardian* (US edition), May 1, 2025, https://www.theguardian.com/environment/2025/may/01/trump-air-climate-pollution-regulation-100-days.

## Chapter 12

**"To be counted in the census":** Alex Wagner, "The Americans Our Government Won't Count," *The New York Times*, March 30, 2018, https://www.nytimes.com/2018/03/30/opinion/sunday/united-states-census.html.

**"largest and most influential American metropolis":** George Lankevich, "New York City," *Encyclopedia Britannica*, January 3, 2025, updated April 6, 2025, https://www.britannica.com/place/New-York-City.

**"No census has ever been taken of the squirrels of Central Park":** "Central Park Squirrels: An Endless Source of Delight to Visitors, Young and Old," *The New York Sun*, October 21, 1900, image 17, https://chroniclingamerica.loc.gov/lccn/sn83030272/1900-10-21/ed-1/seq-17/.

**"The effect of all this upon the squirrels":** "Squirrels and Men," *The Evening World's Daily Magazine*, December 18, 1906, https://www.nyshistoricnewspapers.org/.

**Bisanzio had the science chops:** Donal Bisanzio et al., "Evidence for West Nile Virus Spillover into the Squirrel Population in Atlanta, Georgia," *Vector-Borne and Zoonotic Diseases* 15, no. 5 (May 2015): 303–10, https://doi.org/10.1089/vbz.2014.1734.

**"prioritize native plant species that nourish our wild residents":** Jamie Allen, "Getting to Know Central Park's Squirrels," *Central Park Conservancy Magazine*, October 11, 2022, https://www.centralparknyc.org/articles/getting-to-know-central-parks-squirrels.

**"like stars in a galaxy":** Allen, "Getting to Know Central Park's Squirrels."

**"It's a rectangular green haven in a sea of cement":** Allen, "Getting to Know Central Park's Squirrels."

**"It's important to note that when you engage in any kind of census":** Allen, "Getting to Know Central Park's Squirrels."

**"pedestrian animals":** Allen, "Getting to Know Central Park's Squirrels."

## Chapter 13

**"Darwin's theory of evolution is a framework":** Alan Boyle, "Einstein and Darwin: A Tale of Two Theories," NBC News, April 15, 2005, https://www.nbcnews.com/id/wbna7490426.

**"appears to have been a very rapid divergence":** Monte Basgall, "Squirrels' Evolutionary Family Tree Reveals Influence of Climate, Geology," *Duke Today*, February 20, 2003, https://today.duke.edu/2003/02/squirreltree0302.html.

"By modifying habitats and creating bridges": John M. Mercer and V. Louise Roth, "The Effects of Cenozoic Global Change on Squirrel Phylogeny," *Science* 299, no. 5612 (March 7, 2003): 1568–72, https://doi.org/10.1126/science.1079705.

"The squirrel family (Sciuridae) is one of very few": Mercer and Roth, "Effects of Cenozoic Global Change on Squirrel Phylogeny."

"Look at the family of squirrels": Charles Darwin, "Difficulties on Theory," chap. 6 in *On the Origin of Species by Means of Natural Selection, or, the Preservation of Favoured Races in the Struggle for Life* (John Murray, 1859), 180–81, https://archive.org/details/darwin-online_1859_Origin_F373/page/n197/mode/2up.

"'We're sweating our tails off!'": Hannah Lee, "Maryville College Squirrels Prepare for Winter Amidst Unseasonable Heat: A Humorous Look into Their Minds," *The Highland Echo*, November 19, 2024, http://highlandecho.com/maryville-college-squirrels-prepare-for-winter-amidst-unseasonable-heat-a-humorous-look-into-their-minds/.

"direct study of hunting behavior by squirrels": Eli Wizevich, "Squirrels Are Displaying 'Widespread Carnivorous Behavior' for the First Time in a California Park, New Study Finds," *Smithsonian Magazine*, December 20, 2024, https://www.smithsonianmag.com/smart-news/squirrels-are-displaying-widespread-carnivorous-behavior-for-the-first-time-in-a-california-park-new-study-finds-180985707/.

"ecology, life histories, and physiology": Miyako H. Warrington and Jane Waterman, "Temperature-Associated Morphological Changes in an African Arid-Zone Ground Squirrel," *Journal of Mammalogy* 104, no. 2 (April 2023): 410–20, https://doi.org/10.1093/jmammal/gyac107.

In a 1976–99 study, mountain wagtails: "Animals Shrinking from Climate Change," University of Cape Town, Faculty of Science, July 18, 2019, https://science.uct.ac.za/articles/2019-07-18-animals-shrinking-climate-change; Jorinde Prokosch et al., "Are Animals Shrinking Due to Climate Change? Temperature-Mediated Selection on Body Mass in Mountain Wagtails," *Oecologia* 189 (2019): 841–49, https://doi.org/10.1007/s00442-019-04368-2.

One study found a 0.6 percent body mass decline: Casey Youngflesh et al., "Abiotic Conditions Shape Spatial and Temporal Morphological Variation in North American Birds," *Nature Ecology & Evolution* 6, (2022): 1860–70, https://doi.org/10.1038/s41559-022-01893-x.

"Cape ground squirrels are ecosystem engineers": University of Manitoba, "Squirrel Sperm and Feet Tell a Different Climate Change Story," *Phys.org*, November 29, 2022, https://phys.org/news/2022-11-squirrel-sperm-feet-climate-story.html.

"We saw a pattern": University of Manitoba, "Squirrel Sperm and Feet."

"As species that are able to respond quickly": Warrington and Waterman, "Temperature-Associated Morphological Changes in an African Arid-Zone Ground Squirrel."

**Canadian researchers explored the expansion:** Colin C. Garroway et al., "Climate Change Induced Hybridization in Flying Squirrels," *Global Change Biology* 16, no. 1 (January 2010): 113–21, https://doi.org/10.1111/j.1365-2486.2009.01948.x.

**when the squirrels emerged from hibernation earlier:** Caila E. Kucheravy et al., "Extreme Climate Event Promotes Phenological Mismatch Between Sexes in Hibernating Ground Squirrels," *Scientific Reports* 11 (November 4, 2021): 21684, https://doi.org/10.1038/s41598-021-01214-5.

**female squirrels emerged from their burrows before their male counterparts:** Helen E. Chmura et al., "Climate Change Is Altering the Physiology and Phenology of an Arctic Hibernator," *Science* 380, no. 6647 (2023): 846–49, https://doi.org/10.1126/science.adf5341.

**"Overall, we are interested in using squirrels as indicators of forest change":** Dana Kobilinsky, "TWS Member Gets Up Close with Squirrels," The Wildlife Society, May 22, 2015, https://wildlife.org/tws-member-gets-up-close-and-somewhat-personal-with-squirrels/.

**"In the urban environment there are":** James Dau, "Squirrels and the City," University of Michigan, Rackham Graduate School, April 2, 2020, https://rackham.umich.edu/discover-rackham/squirrels-and-the-city/.

## Chapter 14

**"There are some four million kinds of plants and animals":** David Attenborough, *Life on Earth: A Natural History*, Episode E01, "The Infinite Variety," documentary film, BBC, 1979, https://archive.org/details/life-on-earth-1979/Life+on+Earth+(1979)+-+E01+-+The+Infinite+Variety.mkv.

**"The greatest enemy of the Gray-squirrel":** Ernest Thompson Seton, *Lives of Game Animals*, vol. 4 (Doubleday, Doran & Company, 1929), 51.

**"It's really only here, in the US":** "Sciuridology (SQUIRRELS) with Dr. Karen Munroe," *Ologies with Alie Ward* (podcast), July 25, 2023, https://www.alieward.com/ologies/sciuridology.

**"The only places you find squirrels anymore in Japan":** "Sciuridology (SQUIRRELS) with Dr. Karen Munroe."

**"shortened season of warmer temperatures":** Iral R. Ragenovich and Russel G. Mitchell, "Balsam Woolly Adelgid," US Department of Agriculture, Forest Service, Forest Insect & Disease Leaflet 118, revised May 2006, 5, https://www.michigan.gov/invasives/-/media/Project/Websites/invasives/Documents/ID/Disease/USDA-FS_Balsam_Woolly_Adelgid_Leaflet.pdf.

**about 33 percent of the Nantahala and Pisgah National Forests:** US Fish and Wildlife Service, "Carolina Northern Flying Squirrel (*Glaucomys sabrinus coloratus*) 5-Year Review: Summary and Evaluation," March 2022, https://ecosphere-documents-production-public.s3.amazonaws.com/sams/public_docs/species_nonpublish/3681.pdf.

**"The Service determines endangered status":** US Department of the Interior, Fish and Wildlife Service, "Endangered and Threatened Wildlife and Plants; Determination of Endangered Status for the Mount Graham Red Squirrel," *Federal Register* 52, no. 106 (June 3, 1987): 20994, https://www.govinfo.gov/link/fr/52/20994?link-type=pdf.

**"It's pathetic that the [US] Fish and Wildlife Service recognized":** Center for Biological Diversity, "Lawsuit Seeks Long-Delayed Habitat Expansion for Endangered Mount Graham Red Squirrels in Arizona," press release, March 19, 2024, https://biologicaldiversity.org/w/news/press-releases/lawsuit-seeks-long-delayed-habitat-expansion-for-endangered-mount-graham-red-squirrels-in-arizona-2024-03-19/.

**"This survey process allows us to obtain":** Arizona Game and Fish Department, "Endangered Red Squirrel Making a Huge Comeback," news release, December 10, 2024, https://www.azgfd.com/2024/12/10/endangered-squirrel-making-a-comeback/.

**"They are one of the most challenging species":** Tara Harris, interview by the author, April 23, 2025.

**"Sadly, western gray squirrels are barely hanging on":** "Western Gray Squirrels Granted Washington State Endangered Status: Logging, Climate Change, Sprawl Threaten Rare Forest Squirrel," Center for Biological Diversity, press release, November 17, 2023, https://biologicaldiversity.org/w/news/press-releases/western-gray-squirrels-granted-washington-state-endangered-status-2023-11-17/.

**an indicator species for the "health of rare oak woodland habitat":** Shari Phiel, "Vancouver Group Among Those Suing Washington over Squirrel," *The Columbian*, March 6, 2024, https://www.columbian.com/news/2024/mar/06/vancouver-group-among-those-suing-washington-over-squirrel/.

**"My first year, in 2019, out there we were finding burrows":** Collette Yee, interview by the author via Zoom, May 8, 2024.

**"There are species that the more you get to know":** Yee, interview.

**"I would love to do more research on them":** Yee, interview.

**"People do think that they are almost":** Kelly Wallis, interview by the author via Zoom, August 19, 2024.

**"You are right. Squirrels are very understudied":** Wallis, interview.

**Morris believes that if the researchers discover:** Ralph Bartholdt, "Standing Their Ground: Researcher Studies How Fire Management Affects Threatened Ground Squirrels," *University of Idaho News*, November 2023, https://www.uidaho.edu/news/feature-stories/ground-squirrel.

**"something that human correlative studies":** Irene Torres-Blas et al., "Impact of Exposure to Urban Air Pollution on Grey Squirrel (*Sciurus carolinensis*) Lung Health," *Environmental Pollution* 326 (June 1, 2023): 121312, https://doi.org/10.1016/j.envpol.2023.121312.

"**The crucial point is that animals**": Robin McKie, "Squirrels Live Longer in Leafier Parts of London, Air Pollution Study Shows," *The Guardian* (US edition), April 1, 2023, https://www.theguardian.com/science/2023/apr/01/squirrels-live-longer-in-leafier-parts-of-london-air-pollution-study-shows.

## Chapter 15

"**The time to protect a species**": Author's visit to Hawk Mountain Sanctuary, April 24, 2024.

"**Warm wishes on Squirrel Appreciation Day**": Sumanti Den, "National Squirrel Appreciation Day 2024: All About the January 21 Celebration Dedicated to Our Furry Neighbors," *Hindustan Times*, January 21, 2024, https://www.hindustantimes.com/world-news/us-news/national-squirrel-appreciation-day-2024-all-about-the-january-21-celebration-dedicated-to-our-furry-neighbours-101705803016029.html.

"**But what is the character of our gratitude to these squirrels**": Henry David Thoreau, *Faith in a Seed: The Dispersion of Seeds and Other Late Natural History Writings*, ed. Bradley Dean (Island Press, 1996), 130.

"**suspected there was a deep, untapped reservoir of squirrel love**": John Kelly, "When Skwerls Attack," *The Washington Post*, March 17, 2011.

"**We don't keep track of the number of squirrels taken**": Eric Maynard, interview by the author.

"**We should understand well that all things are the works of the Great Spirit**": Black Elk, *The Sacred Pipe: Black Elk's Account of the Seven Rites of the Oglala Sioux*, recorded and edited by Joseph Epes Brown (University of Oklahoma Press, 1953), xx.

**a more extensive nasal epithelium**: Karen K. Yee et al., "Comparative Morphology and Histology of the Nasal Fossa in Four Mammals: Gray Squirrel, Bobcat, Coyote, and White-Tailed Deer," *The Anatomical Record* 299, no. 7 (July 2016): 840–52, https://doi.org/10.1002/ar.23352.

**it had taken years to get the squirrels to this ability level**: Lyric Li and Annabelle Timsit, "A Squad of Drug-Sniffing Squirrels Is Training to Join China's Police," *The Washington Post*, February 9, 2023, https://www.washingtonpost.com/world/2023/02/09/squirrels-drugs-police-china-chongqing/.

"**She has since taken a desk job**": Anna Sanders, "Meet the Hidden—Sometimes Bizarre—Mascots of NYC," *New York Post*, December 30, 2017, https://nypost.com/2017/12/30/meet-the-hidden-sometimes-bizarre-mascots-of-nyc/.

"**Should we preserve the Red-squirrels**": Ernest Thompson Seton, *Life-Histories of Northern Animals: An Account of the Mammals of Manitoba* (Charles Scribner's Sons, 1909), 336.

"**My experience is that if you tap on about 1,000 trees**": Thom Smith, "Flying Squirrels May Not Be as Numerous as Gray Squirrels, but They're Holding Their Own in the Berkshires," *The Berkshire Eagle*, February 16, 2022, https://www.berkshireag

le.com/arts_and_culture/home-garden/flying-squirrels-juncos-tax-donations-thom-smith-naturewatch/article_b2a1ed66-8dea-11ec-8679-eb9ea0d54331.html.

**just 3 percent of land-dwelling mammals met their end on a highway:** Ben Goldfarb, *Crossings: How Road Ecology Is Shaping the Future of Our Planet* (W. W. Norton, 2023), 4.

**urban environmental characteristics are stressors for squirrels:** Pizza Ka Yee Chow et al., "Characteristics of Urban Environments and Novel Problem-Solving Performance in Eurasian Red Squirrels," *Proceedings of the Royal Society B: Biological Sciences* 288, no. 1947 (March 2021): 20202832, https://doi.org/10.1098/rspb.2020.2832.

**"Squirrels help us study how ecosystems are changing":** Dana Kobilinsky, "TWS Member Gets Up Close with Squirrels," The Wildlife Society, May 22, 2015, https://wildlife.org/tws-member-gets-up-close-and-somewhat-personal-with-squirrels/.

**there is sometimes good news:** Emiliano Mori and Mattia Menchetti, "'Sometimes They Come Back': Citizen Science Reveals the Presence of the Italian Red Squirrel in Campania," *Quaderni del Museo di Storia Naturale di Ferrara* 2 (2014): 91–94, https://www.researchgate.net/publication/268388222_Sometimes_they_come_back_citizen_science_reveals_the_presence_of_the_Italian_red_squirrel_in_Campania.

**"What does it do to children":** Etienne Benson, interview by the author, April 2024.

# *Bibliography*

Ackerman, Diane. *Cultivating Delight: A Natural History of My Garden*. G. K. Hall, 2002.
Ackerman, Diane. "In Praise of Squirrels." *National Geographic* 188, no. 5 (November 1995).
Alagona, Peter S. *The Accidental Ecosystem*. University of California Press, 2022.
*A Non-Lethal Management Guide for Gunnison's Prairie Dogs*. Habitat Harmony, 2018. https://habitatharmony.org/handbook.
Arizona Game and Fish Department. "Endangered Red Squirrel Making a Huge Comeback." December 10, 2024. https://www.azgfd.com/2024/12/10/endangered-squirrel-making-a-comeback/.
Babuji, Varsha Sara. "Heroic Mongoose Rescues Squirrel from Cobra in Viral Video—Watch." *News9 Live*, February 29, 2024. https://www.news9live.com/videos/viral-videos/heroic-mongoose-rescues-squirrel-from-cobra-in-viral-video-watch-2453644.
Badry, Alexander, Detlef Schenke, Gabriele Treu, and Oliver Krone. "Linking Landscape Composition and Biological Factors with Exposure Levels of Rodenticides and Agrochemicals in Avian Apex Predators from Germany." *Environmental Research* 193 (February 2021): 110602. https://doi.org/10.1016/j.envres.2020.110602.
Bailey, Vernon. "Mammals of the Vicinity of Washington." *Journal of the Washington Academy of Sciences* 16, no. 16 (October 4, 1926): 441–45. https://www.jstor.org/stable/24529385.
Balakrishna, Udbhavi. "Bengaluru Sculptor's Symbolic Squirrel Statue to Greet Passengers at Ayodhya Station." *Deccan Herald*, January 10, 2024. https://www.deccanherald.com/india/karnataka/bengaluru-city-sculptor-s-symbolic-squirrel-statue-to-greet-passengers-at-ayodhya-station-2843857.
Barkalow, Frederick S., Jr., and Monica Shorten. *The World of the Gray Squirrel*. Lippincott Williams & Wilkins, 1973.
Bartram, William. *The Travels of William Bartram*. Macy-Masius, 1928. https://books.google.com/books/about/The_Travels_of_William_Bartram.html?id=-u1HAAAAIAAJ.
Benson, Etienne. "The Urbanization of the Eastern Gray Squirrel in the United States." *The Journal of American History* 100, no. 3 (2013): 691–710. http://www.jstor.org/stable/44308759.
Berger, Michael. "Henry David Thoreau's Science in *The Dispersion of Seeds*." *Annals of Science* 53, no. 4 (1996): 381–97. https://doi.org/10.1080/00033799608560823.
Bernstein, Jules. "How Gophers Brought Mount St. Helens Back to Life in One Day." University of California, November 7, 2024. https://www.universityofcalifornia.edu/news/how-gophers-brought-mount-st-helens-back-life-one-day.

Bertolino, Sandro. "Animal Trade and Non-Indigenous Species Introduction: The World-Wide Spread of Squirrels." *Diversity and Distributions* 15, no. 4 (June 10, 2009): 701–8. https://doi.org/10.1111/j.1472-4642.2009.00574.x.

Bertolino, Sandro, Italo Currado, Peter Mazzoglio, and Giovanni Amori. "Native and Alien Squirrels in Italy." *Hystrix, the Italian Journal of Mammalogy* 11, no. 2 (2000): 65–74. https://doi.org/10.4404/hystrix-11.2-4150.

Bertrand, Ornella C., Hans P. Püschel, Julia A. Schwab, Mary T. Silcox, and Stephen L. Brusatte. "The Impact of Locomotion on the Brain Evolution of Squirrels and Close Relatives." *Communications Biology* 4, art. 460 (2021). https://doi.org/10.1038/s42003-021-01887-8.

Black Elk. *The Sacred Pipe: Black Elk's Account of the Seven Rites of the Oglala Sioux*, recorded and edited by Joseph Epes Brown. University of Oklahoma Press, 1953.

Bodin, Madeline. "Cordie Diggins' Research Featured in *Nature Conservancy Magazine*." Virginia Tech, Global Change Center, February 10, 2015. https://globalchange.vt.edu/news/news-stories/2014-15-news/diggins-featured-in-nature-conservancy-magazine.html.

Boone, Wesley W., Brittany A. Bankovich, Brian E. Reichert, Mandy B. Watson, and Robert A. McCleery. "Frequent Prescribed Burns Reduce Mammalian Species Richness and Occurrence in Longleaf Pine Sandhills." *Forest Ecology and Management* 553 (February 1, 2024): 121596. https://doi.org/10.1016/j.foreco.2023.121596.

Brickell, John. *The Natural History of North Carolina: With an Account of the Trade, Manners, and Customs of the Christian and Indian Inhabitants. Illustrated with a Map, and Copper-Plates, etc.* Charles Corbett, 1743.

Brown, Alex. "No More Prizes for Killing 'Nuisance' Animals Under These Hunting Contest Bans." *New Hampshire Bulletin*, January 30, 2024. https://newhampshirebulletin.com/2024/01/30/no-more-prizes-for-killing-nuisance-animals-under-these-hunting-contest-bans/.

"The Brunswick Stew: The Originator of It and How the Dish Was First Made." *National Republican*, September 16, 1886, Image 3. https://chroniclingamerica.loc.gov/lccn/sn86053573/1886-09-16/ed-1/seq-3/.

Burroughs, John. *Squirrels and Other Fur-Bearers*. Houghton Mifflin, 1901.

Carini, Frank, and Colleen Cronin. "The Secret Is Out: Some Local Squirrels Can Fly." *ecoRI News*, January 5, 2024. https://ecori.org/the-secret-is-out-some-local-squirrels-can-fly/.

Carlen, Elizabeth J., Cesar O. Estien, Tal Caspi, et al. "A Framework for Contextualizing Social-Ecological Biases in Contributory Science Data." *People and Nature* 6, no. 2 (March 3, 2024): 377–90. https://doi.org/10.1002/pan3.10592.

Carson, Rachel. *Silent Spring*. Houghton Mifflin, 1962.

Chow, Pizza Ka Yee, Nicola S. Clayton, and Michael A. Steele. "Cognitive Performance of Wild Eastern Gray Squirrels (*Sciurus carolinensis*) in Rural and Urban, Native, and Non-Native Environments." *Frontiers in Ecology and Evolution* 9 (February 26, 2021). https://doi.org/10.3389/fevo.2021.615899.

Coates, Peter A. *Squirrel Nation: Reds, Greys and the Meaning of Home*. Reaktion Books, 2023.

Daley, Jason. "Volunteers Counted All the Squirrels in Central Park." *Smithsonian*

*Magazine*, June 24, 2019. https://www.smithsonianmag.com/smart-news/new-cen sus-counted-all-squirrels-central-park-180972480/.

Davis, Jerry. "Maple Sap Attracts Naturalists." *Ag Update*, February 17, 2024. https:// agupdate.com/agriview/lifestyles/maple-sap-attracts-naturalists/article_ac7c4fdc -ca95-11ee-8671-6f066bf4ab35.html.

Delaware Tribe of Indians. "Welcome to the Lenape Talking Dictionary." *Lenape Talking Dictionary*, accessed 2024. https://www.talk-lenape.org/.

DiCamillo, Kate. *Flora and Ulysses: The Illuminated Adventures*. Candlewick Press, 2013.

Dittel, Jacob W., Ramón Perea, and Stephen B. Vander Wall. "Reciprocal Pilfering in a Seed-Caching Rodent Community: Implications for Species Coexistence." *Behavioral Ecology and Sociobiology* 71, no. 147 (2017): 1–10. https://doi.org/10.1007 /s00265-017-2375-4.

Du, Susan. "Did a Famed Parks Leader Import Gray Squirrels to Minneapolis—and Have the Red Ones Killed?" *Minnesota Star Tribune*, January 20, 2023. https://ww w2.startribune.com/gray-red-squirrels-theodore-wirth-loring-park-birds-minnea polis/600245052/.

Eder, Klaus. *The Social Construction of Nature: A Sociology of Ecological Enlightenment*. Translated by M. Ritter. Sage, 1996.

Emry, Robert J., and William W. Korth. "A New Genus of Squirrel (Rodentia, Sciuridae) from the Mid-Cenozoic of North America." *Journal of Vertebrate Paleontology* 27, no. 3 (2007): 693–98. https://doi.org/10.1671/0272-4634(2007)27[693:ANGOSR ]2.0.CO;2.

Emry, Robert J., and Richard W. Thorington Jr. "Descriptive and Comparative Osteology of the Oldest Fossil Squirrel, *Protosciurus* (Rodentia: Sciuridae)." *Smithsonian Contributions to Paleobiology*, no. 47. Smithsonian Institution Press, 1982. https://re pository.si.edu/bitstream/handle/10088/19142/SCTP-0047.pdf.

Faber, Nicky R., Gus R. McFarlane, R. Chris Gaynor, Ivan Pocrnic, C. Bruce A. Whitelaw, and Gregor Gorjanc. "Novel Combination of CRISPR-Based Gene Drives Eliminates Resistance and Localises Spread." *Scientific Reports* 11, art. 3719 (2021). https://doi.org/10.1038/s41598-021-83239-4.

"Fauna." *Parchi Reali Torino*, December 17, 2024. https://parchireali.it/parco-naturale -di-stupinigi/fauna/.

Flores, Dan. *Wild New World: The Epic Story of Animals and People in America* (W. W. Norton, 2022).

Forest Preserve District of Will County, Illinois. "Five Fun Facts About Our Squirrelly Squirrels," January 3, 2024. https://www.reconnectwithnature.org/news-events/the -buzz/five-fun-facts-you-need-to-know-about-squirrels/.

Forgrave, Andrew. "Cheeky Squirrel Spotted on Snowdonia Peak Looking for Easy Pickings from Walkers." *North Wales Live*, August 3, 2023. https://www.dailypost.co .uk/news/north-wales-news/cheeky-squirrel-spotted-snowdonia-peak-27449773.

Franklin, Benjamin. "From Benjamin Franklin to Georgiana Shipley, 26 September 1772." National Archives and Records Administration, *Founders Online*. Accessed January 15, 2024. https://founders.archives.gov/documents/Franklin/01-19-02 -0202.

"Friendly Indian Giant Squirrel Welcomes Visitors at the Adavi Eco Tourism Centre."

*Onmanorama*, May 2, 2023. https://www.onmanorama.com/travel/kerala/2023/04/30/malayannaan-indian-giant-squirrel-adavi-eco-tourism-centre.html.

Gallo, Orlando, Arnaldo Iudici, and Rosario Balestrieri. "Farther Inland Invasion of Finlayson's Squirrel *Callosciurus finlaysonii* (Horsfield 1823) Poses a New Conservation Challenge for the Endemic Near Threatened Calabrian Black Squirrel *Sciurus meridionalis* Lucifero 1907 (Rodentia: Sciuridae)." *Natural History Sciences* 11, no. 2 (2024). https://doi.org/10.4081/nhs.2024.760.

Genoa City Tours. "Nervi Parks—Serra Gropallo—Nervi." Accessed January 2, 2025. https://www.genoa.in/genoa/what-to-see/details/parchi-di-nervi-serra-gropallo.

Genysys Engine. "Squirrel Agent: Project Red Haven." Accessed November 25, 2024. https://genysysengine.tech/squirrel-feeding-agent.

Gilliland, Steve. "Exploring Kansas Outdoors: Squirrels; Love 'Em or Hate 'Em." *Hays Post*, February 11, 2024. https://hayspost.com/posts/16895b94-700a-4729-ac29-8eb149aa048b.

Godman, John Davidson. *American Natural History*. Vol. 2. Carey, Lea & Carey, 1828.

Goldfarb, Ben. *Crossings: How Road Ecology Is Shaping the Future of Our Planet*. W. W. Norton, 2023.

Grenrock, Samantha. "Why Should You Love Squirrels? Here Are Six Reasons." University of Florida News Archive, January 17, 2018. https://news.ufl.edu/articles/2018/01/why-should-you-love-squirrels-here-are-six-reasons.html.

Grindon, Lucy. "One Last Rabbit Hunt in Norfolk Before Some Hunting Competitions Are Banned." NCPR, January 23, 2024. https://www.northcountrypublicradio.org/news/story/49121/20240123/one-last-rabbit-hunt-in-norfolk-before-some-hunting-competitions-are-banned.

Guerrero, Carmen. "Celebrating Fat Squirrel Week: Squirrel Behavior on Campus and Why We Shouldn't Feed Them." Colorado State University, Warner College of Natural Resources, October 8, 2024. https://warnercnr.source.colostate.edu/celebrating-fat-squirrel-week-squirrel-behavior-on-campus-and-why-we-shouldnt-feed-them/.

Hatt, Robert T. "The Red Squirrel: Its Life History and Habits, with Special Reference to the Adirondacks of New York and the Harvard Forest." *Roosevelt Wild Life Annals* 2, no. 1 (March 1929). Roosevelt Wild Life Forest Experiment Station.

Heinrich, Bernd. "Maple Sugaring by Red Squirrels." *Journal of Mammalogy* 73, no. 1 (April 14, 1992): 51–54. https://doi.org/10.2307/1381865.

Hoage, Natalie. "Man Sets Up Most Elaborate House for the Wild Squirrels in His Neighborhood." *Yahoo!*, January 23, 2024. https://www.yahoo.com/lifestyle/man-sets-most-elaborate-house-143153450.html.

Houghton, Amy. "Rare (and Very Cute) Baby Red Squirrels Have Been Born in Yorkshire." *Time Out United Kingdom*, August 11, 2023. https://www.timeout.com/uk/news/rare-and-very-cute-baby-red-squirrels-have-been-born-in-yorkshire-081123.

Hughes, Seren. "Four Rare Red Squirrels Born in Yorkshire Enclosure." *The London Times*, July 28, 2023. https://www.thetimes.co.uk/article/four-rare-red-squirrels-born-in-yorkshire-enclosure-mgv2obtr5.

Jaffe, Matt. "Treated like a Dog!" *Arizona Highways*. Accessed November 12, 2024. https://www.arizonahighways.com/article/treated-dog.

Jewell, Susan D. "A Century of Injurious Wildlife Listing Under the Lacey Act: A History."

*Management of Biological Invasions* 11, no. 3 (September 2020): 356–71. https://doi.org/10.3391/mbi.2020.11.3.01.

Johansson, Maria, Terry Hartig, Jens Frank, and Anders Flykt. "Wildlife and Public Perceptions of Opportunities for Psychological Restoration in Local Natural Settings." *People and Nature* 6, no. 2 (April 2024): 800–817. https://doi.org/10.1002/pan3.10616.

Jones, Terry L. "Alligator Tales: Stories from Louisiana's Alligator-Filled Past." *Louisiana Sportsman*, September 1, 2012. https://www.louisianasportsman.com/hunting/other-hunting/alligator-tales-stories-from-louisianas-alligator-filled-past/.

Jones, Terry L. "A Fall Tradition." *Country Roads Magazine*, September 23, 2019. https://countryroadsmagazine.com/outdoors/Squirrel-season-memories-pastimes/.

Juutinen, Artti, Suvi Ilvonen, Emmi Haltia, et al. "Citizens' Attitudes Toward the Protection of Flying Squirrels in Urban Areas." *Ecology & Society* 28, no. 4 (December 26, 2023): art. 19. https://doi.org/10.5751/ES-14190-280419.

Kelly, John. "Squirrel Camp: Meet the Scientists Who Live off the Grid to Study the Frisky Critters." *The Washington Post*, April 13, 2019. https://www.washingtonpost.com/local/squirrel-camp-meet-the-scientists-who-live-off-the-grid-to-study-the-frisky-critters/2019/04/13/a764102c-5bb2-11e9-a00e-050dc7b82693_story.html.

Kimmerer, Robin Wall. "Maple Sugar Moon." In *Braiding Sweetgrass: Indigenous Wisdom, Scientific Knowledge, and the Teachings of Plants*, 63–71. Milkweed Editions, 2015.

Klingaman, Gerald. "Squirrels." *Musings on Nature: Out and About with Gerald & Whinny*. University of Arkansas Division of Agriculture Research & Extension. Accessed February 26, 2024. https://www.uaex.uada.edu/environment-nature/musings-on-nature-blog/posts/squirrels.aspx.

Lawrence, D. H. "To Nuoro." Chap. 6 in *Sea and Sardinia*. Thomas Seltzer, 1921. http://www.online-literature.com/dh_lawrence/sea-and-sardinia/6/.

Lee, Hannah. "Maryville College Squirrels Prepare for Winter Amidst Unseasonable Heat: A Humorous Look into Their Minds." *The Highland Echo*, November 19, 2024. http://highlandecho.com/maryville-college-squirrels-prepare-for-winter-amidst-unseasonable-heat-a-humorous-look-into-their-minds/.

Leibniz Institute for Zoo and Wildlife Research (IZW). "Rodenticides in the Environment Pose Threats to Birds of Prey." *ScienceDaily*, March 24, 2021. https://www.sciencedaily.com/releases/2021/03/210324135430.htm.

Leopold, Aldo. *Round River: From the Journals of Aldo Leopold*. Oxford University Press, 1993.

MacLoughlin, Shaun. "English Wordplay—Listen and Enjoy: Percy Bysshe Shelley, Journey to Italy and *The Cenci*." *English Wordplay*, accessed November 9, 2024. http://www.englishwordplay.com/shelley6.html.

Maltz, Mia Rose, Michael F. Allen, Michala L. Phillips, et al. "Microbial Community Structure in Recovering Forests of Mount St. Helens." *Frontiers in Microbiomes* 3 (November 3, 2024). https://doi.org/10.3389/frmbi.2024.1399416.

"Mammals of the Adirondacks: Red Squirrel (*Tamiasciurus hudsonicus*)." *Wild Adirondacks*. Accessed November 20, 2024. https://wildadirondacks.org/adirondack-mammals-red-squirrel-tamiasciurus-hudsonicus.html.

Mazzamuto, Maria Vittoria, Francesco Bisi, Lucas A. Wauters, Damiano G. Preatoni, and Adriano Martinoli. "Interspecific Competition Between Alien Pallas's Squirrels and Eurasian Red Squirrels Reduces Density of the Native Species." *Biological Invasions* 19, no. 2 (November 3, 2016): 723–35. https://doi.org/10.1007/s10530-016-1310-3.

McRae, Thaddeus R. "A Review of Squirrel Alarm-Calling Behavior: What We Know and What We Do Not Know About How Predator Attributes Affect Alarm Calls." *Animal Behavior and Cognition* 7, no. 2 (2020): 168–91. https://doi.org/10.26451/abc.07.02.11.2020.

Mendes, Sara Beatriz, Jens Mogens Olesen, Jane Memmott, et al. "Evidence of a European Seed Dispersal Crisis." *Science* 386, no. 6718 (October 10, 2024): 206–11. https://doi.org/10.1126/science.ado1464.

Mohan, Vishnu. "'Not a Fan of People Who Kill for Sport'—When Brian Harman Opened Up About Hunting and the Life Lessons It Gave." *Sportskeeda*, July 24, 2023. https://www.sportskeeda.com/golf/news-not-fan-people-kill-sport-when-brian-harman-opened-hunting-life-lessons-gave.

Morgan, Kate. "How (and Why) to Peacefully Coexist with Squirrels." *The Washington Post*, February 21, 2024. https://www.washingtonpost.com/home/2024/02/22/living-with-squirrels-nuisance/.

Mori, Emiliano, and Mattia Menchetti. "'Sometimes They Come Back': Citizen Science Reveals the Presence of the Italian Red Squirrel in Campania." *Quaderni del Museo di Storia Naturale di Ferrara* 2 (2014): 91–94. https://www.researchgate.net/publication/268388222_Sometimes_they_come_back_citizen_science_reveals_the_presence_of_the_Italian_red_squirrel_in_Campania.

Morrison, Ashley. "Sparks Fly at the Third Annual Piedmont Squirrel Drop." *The Anniston Star*, January 2, 2024. https://www.annistonstar.com/news/piedmont/sparks-fly-at-the-third-annual-piedmont-squirrel-drop/article_2d0d2958-a9c3-11ee-8cb2-07b230a0947b.html.

Morrissey, Laurie D. "Squirrel Talk: What Does That Noise Mean?" *Northern Woodlands*, November 28, 2022. https://northernwoodlands.org/outside_story/article/squirrel-talk.

Mussulman, Joseph A. "Eastern Gray Squirrel." *Discover Lewis & Clark*, February 20, 2022. https://lewis-clark.org/sciences/mammals/eastern-gray-squirrel/.

Nair, Anirudh. "Hop, Skip and Jump: The Giant Squirrels of India." *Roundglass Sustain*, January 20, 2021, updated September 24, 2023. https://roundglasssustain.com/photo-stories/giant-squirrels-of-india.

Nassauer, Sarah. "Welcome Back, Wal-Mart Greeters." *The Wall Street Journal*, June 18, 2015. https://www.wsj.com/articles/wal-mart-ushers-greeters-back-to-the-front-1434651944.

*NCPR News*. "1/23/24: Goodbye, Rabbit and Squirrel Hunting Tournaments." NCPR, January 23, 2024. https://www.northcountrypublicradio.org/news/story/49162/20240123/1-23-24-goodbye-rabbit-and-squirrel-hunting-tournaments.

*New-York Daily Tribune*. "Willcox Squirrels' Friend: The New Commissioner Provides Peanuts for Hungry Park Dwellers." January 11, 1902. https://chroniclingamerica.loc.gov/lccn/sn83030214/1902-01-11/ed-1/seq-10.

Nosowitz, Dan. "Is a Squirrel Smarter than a Fifth-Grader?" *Atlas Obscura*, August 13, 2015. https://www.atlasobscura.com/articles/is-a-squirrel-smarter-than-a-fifth grader.

O'Brien, Haley. "The Search for Endangered Flying Squirrels in the Poconos." *The Allegheny Front*, February 16, 2024, updated March 6, 2024. https://www.allegheny front.org/pennsylvania-poconos-endangered-northern-flying-squirrels/.

Ogliore, Talia. "How Bias Shows Up in Maps Made with Citizen Science Data." Washington University in St. Louis, *The Source*, March 5, 2024. https://source.was hu.edu/2024/03/how-bias-shows-up-in-maps-made-with-citizen-science-data/.

"Palazzina di Caccia di Stupinigi." Fondazione Ordine Mauriziano, September 9, 2024. https://www.ordinemauriziano.it/en/palazzina-di-caccia-di-stupinigi/.

"Parco di Stupinigi: Protected Area." Accessed January 2, 2025. https://www.parks.it /parco.stupinigi/Epar.php.

*Philadelphia Public Ledger*, May 11, 1847.

Platt, John. "Italy Faces Invasion of American Killer Squirrels." *Scientific American*, October 5, 2012. https://www.scientificamerican.com/blog/extinction-countdown /italy-faces-invasion-of-american-killer-squirrels/.

Pliny the Elder. In *Natural History*, Book 8, 58. *The Medieval Bestiary*. Accessed 2023. https://bestiary.ca/beasts/beastsource105918.htm.

Posthumus, Erin E., John L. Koprowski, and Robert J. Steidl. "Red Squirrel Middens Influence Abundance but Not Diversity of Other Vertebrates." *PloS One* 10, no. 4 (April 29, 2015): e0123633. https://doi.org/10.1371/journal.pone.0123633.

Potter, Beatrix. *The Complete Tales of Beatrix Potter*. F. Warne & Co., 1989.

"Project Squirrel." University of New England, School of Marine and Environmental Programs, Perlut Lab. Accessed March 4, 2024. https://sites.une.edu/perlutlab /project-squirrel/.

Prokosch, Jorinde, Zephne Bernitz, Herman Bernitz, Birgit Erni, and Res Altwegg. "Are Animals Shrinking Due to Climate Change? Temperature-Mediated Selection on Body Mass in Mountain Wagtails." *Oecologia* 189 (2019): 841–49. https://doi .org/10.1007/s00442-019-04368-2.

Raso, Caterina, Valentina Galietta, Claudia Eleni, et al. "Ectopic Pregnancy and T-Cell Lymphoma in a Eurasian Red Squirrel (*Sciurus vulgaris*): Possible Comorbidity and a Comparative Pathology Perspective." *Animals* 14, no. 5 (2024): 731. https://doi.org/10.3390/ani14050731.

Red Squirrel Survival Trust. "Red Squirrel Vs Grey Squirrel: The Key Differences." *Discover Wildlife*, January 10, 2023. https://www.discoverwildlife.com/animal-facts /mammals/red-squirrel-vs-grey-squirrel-the-key-differences.

Roper, Eric. "Listen: Did a Famed Parks Leader Import Gray Squirrels to Minneapolis—and Have the Red Ones Killed?" *The Minnesota Star Tribune*, April 21, 2023. https://www.startribune.com/curious-minnesota-podcast-gray-red-squirrels /600269080/.

Rozell, Ned. "The Secret Life of Red Squirrels." University of Alaska Fairbanks, *UAF News and Information*, November 27, 2024. https://www.uaf.edu/news/the-secret -life-of-red-squirrels.php.

Rusin, Mikhail. "Threats of Russian Invasion for Protected Small Mammals in

Ukraine." Ukraine War Environmental Consequences Work Group, February 8, 2023. https://uwecworkgroup.info/threats-of-russian-invasion-for-protected-small-mammals-in-ukraine/.

"Sciuridology (SQUIRRELS) with Dr. Karen Munroe." *Ologies with Alie Ward* (podcast), episode 335, July 25, 2023, https://www.alieward.com/ologies/sciuridology.

Serrao, Nivea. "Shannon and Dean Hale on the Unbeatable Experience of Writing *Squirrel Girl*." *Entertainment Weekly*, February 7, 2017. https://ew.com/books/2017/02/07/shannon-hale-dean-hale-unbeatable-squirrel-girl-squirrel-meets-world/.

Seton, Ernest Thompson. *Bannertail: The Story of a Graysquirrel*. Charles Scribner's Sons, 1922.

Signorile, A. L., D. C. Reuman, P. W. W. Lurz, S. Bertolino, C. Carbone, and J. Wang. "Using DNA Profiling to Investigate Human-Mediated Translocations of an Invasive Species." *Biological Conservation* 195 (March 2016): 97–105. https://doi.org/10.1016/j.biocon.2015.12.026.

Silver, Robin, and Charles Babbitt. "Lawsuit Seeks Long-Delayed Habitat Expansion for Endangered Mount Graham Red Squirrels in Arizona." Press release, Center for Biological Diversity, March 19, 2024. https://biologicaldiversity.org/w/news/press-releases/lawsuit-seeks-long-delayed-habitat-expansion-for-endangered-mount-graham-red-squirrels-in-arizona-2024-03-19/.

Smallwood, Peter D., and Wm. David Peters. "Grey Squirrel Food Preferences: The Effects of Tannin and Fat Concentration." *Ecology* 67, no. 1 (February 1986): 168–74. https://doi.org/10.2307/1938515.

Spieler, Marlena, and *The New York Times*. "Squirrel Finds New Popularity Among British Diners." *The Mercury News*, February 3, 2009. https://www.mercurynews.com/2009/02/03/squirrel-finds-new-popularity-among-british-diners/.

"Squirrels and Men." *The Evening World's Daily Magazine*, December 18, 1906. https://www.nyshistoricnewspapers.org/.

"Squirrely (adj.)." *Online Etymology Dictionary*. Accessed January 7, 2024. https://www.etymonline.com/word/squirrely.

Stackhouse, Sean. "Project Squirrel: UNE Students Studying Squirrels Find Unexpected Data." *News Center Maine*, May 1, 2023. https://www.newscentermaine.com/article/life/animals/university-new-england-biddeford-project-squirrel/97-23014 00a-6710-4169-94a7-28e1de8374d5.

Steele, Michael A., Sylvia L. Halkin, Peter D. Smallwood, Thomas J. McKenna, Katerina Mitsopoulos, and Matthew Beam. "Cache Protection Strategies of a Scatter-Hoarding Rodent: Do Tree Squirrels Engage in Behavioural Deception?" *Animal Behaviour* 75, no. 2 (February 2008): 705–14. https://doi.org/10.1016/j.anbehav.2007.07.026.

Steele, Michael A., and John L. Koprowski. *North American Tree Squirrels*. Smithsonian Institution Press, 2003.

Stein, Sadie. "Alien Squirrel." *New York Magazine*, February 3, 2014. https://nymag.com/news/features/squirrels-2014-2/.

Thampuran, Athmaja Varma. "Forest Department Envisages Creating Canopy Bridges for Squirrel Conservation." *Onmanorama*, January 22, 2024. https://www.onmanorama.com/lifestyle/news/2024/01/22/forest-department-envisages-creating-canopy-bridges-squirrel-conservation.html.

Thoreau, Henry David. *Faith in a Seed: The Dispersion of Seeds and Other Late Natural History Writings.* Edited by Bradley Dean. Island Press, 1996.
Thorington, Richard W., Jr., and Katie Ferrell. *Squirrels: The Animal Answer Guide.* John Hopkins University Press, 2006.
University of Surrey. "Scientists Are Unravelling the Secrets of Red and Grey Squirrel Competition." Press release, February 15, 2024. https://www.surrey.ac.uk/news/scientists-are-unravelling-secrets-red-and-grey-squirrel-competition.
Vanek, John P., James P. Gibbs, and Bradley Cosentino. "Melanism in the Eastern Gray Squirrel: Using Multiple Methods to Quantify an Urban-Rural Cline." Poster, April 2022. https://doi.org/10.13140/RG.2.2.30903.60320.
Veigel, Steven N. "Why Millions of Gray Squirrels Crossed America in Swarms." *AtCharlie Chronicles,* January 13, 2023. https://www.atcharlie.com/atcharlie-chronicles/why-millions-of-gray-squirrels-crossed-america-in-swarms/.
Vincent Wildlife Trust Ireland. "8th International Colloquium on Squirrels," accessed November 23, 2024. https://www.vincentwildlife.ie/news-media/blog/8th-international-colloquium-on-squirrels.
Vogel, Steven. "The Social Construction of Nature." In *Thinking Like a Mall: Environmental Philosophy After the End of Nature,* 33–64. MIT Press, 2015. https://doi.org/10.7551/mitpress/9780262029100.003.0002.
"Warren Harding's Pete." Presidential Pet Museum. Accessed February 26, 2024. https://www.presidentialpetmuseum.com/warren-hardings-pete/.
Wells-Gosling, Nancy. Essay in *Flying Squirrels: Gliders in the Dark,* 87. Smithsonian Institution Press, 1985.
Williams, Linda. "Oak Branch Tips Laying on the Ground This Fall." *Wisconsin DNR Forestry News,* September 26, 2018. https://forestrynews.blogs.govdelivery.com/2018/09/26/oak-branch-tips-laying-on-the-ground-this-fall/.
Yahnke, Christopher J. "Xylem Sap Moon." Squirrel-Net, April 19, 2023. https://www.squirrel-net.org/post/xylem-sap-moon.
Yolen, Jane. *Trash Mountain.* Lerner, 2015.
Youngflesh, Casey, James F. Saracco, Rodney B. Siegel, and Morgan W. Tingley. "Abiotic Conditions Shape Spatial and Temporal Morphological Variation in North American Birds." *Nature Ecology & Evolution* 6 (2022): 1860–70. https://doi.org/10.1038/s41559-022-01893-x.

## *Illustration Credits*

**Frontispiece:** *Eastern Gray Squirrel*, John James Audubon. The Morgan Library & Museum. https://www.themorgan.org/drawings/item/143030.

**Preface:** *Common Squirrel* (Eurasian Red Squirrel), 1799. Published by W. Darton, J. Harvey, and W. Belch, London. Author's collection.

**Chapter 1:** Squirrel from *Descrizioni Degli Animali* (1771-1775) by Innocente Alessandri, Pietro Scattaglia and Lodovico Leschi.

**Chapter 2:** Author's collection.

**Chapter 3:** *Red Squirrel*, Hans Hoffman, 1578. Courtesy of the National Gallery of Art, Woodner Collection. https://www.nga.gov/collection/art-object-page.73845.html.

**Chapter 4:** Samuel Howitt, *Flying Squirrel*, ca. 1817, Yale Center for British Art, https://collections.britishart.yale.edu/catalog/tms:10541.

**Chapter 5:** Migratory gray squirrel, *Sciurus migratorius*, 1872. Library of Congress. https://www.loc.gov/item/2017660730/.

**Chapter 6:** *Feeding Squirrels in Union Square*. Cooper Hewitt, Smithsonian Design Museum. Gift of Mrs. George A. Kubler 1882.

**Chapter 7:** *American Red Squirrel* by German School, undated.

**Chapter 8:** *Squirrels*. The Miriam and Ira D. Wallach Division of Art, Prints and Photographs: Print Collection, The New York Public Library. New York Public Library Digital Collections. 1794–1870. Accessed February 6, 2025. https://digitalcollections.nypl.org/items/510d47dc-5ed5-a3d9-e040-e00a18064a99.

**Chapter 9:** *Finlayson's squirrel* from Die Säugthiere in Abbildungen nach der Natur (1778-1855) by Georg August Goldfuss, Johann Andreas Wagener and Johann Christian Daniel von Schreber.

**Chapter 10:** Author's collection.

**Chapter 11:** Author's collection.

**Chapter 12:** Author's collection.

**Chapter 13:** Page tab XCVII. Scan of 276. Meyer, Johann Daniel (1713–1752), 1748. Angenehmer und nützlicher Zeit-Vertreib mit Betrachtung curioser Vorstellungen allerhand kriechender, fliegender und schwimmender [...] Thiere, sowohl nach ihrer Gestalt und äusserlichen Beschaffenheit als auch der accuratest davon verfertigsten Structur ihrer Scelete oder Bein-Cörper nebst einer [...]. Grafika i Rysunek. Courtesy Biblioteka Narodowa.

**Chapter 14:** Author's collection.

**Chapter 15:** *Squirrel Eating Chestnuts*, Kawabata Gyokushō, 1887–92. The Metropolitan Museum of Art. Charles Stewart Smith Collection, Gift of Mrs. Charles Stewart Smith, Charles Stewart Smith Jr., and Howard Caswell Smith, in memory of Charles Stewart Smith, 1914. https://www.metmuseum.org/art/collection/search/54750.

## *About the Author*

Nancy Castaldo is a naturalist, an environmental educator, and the author of more than two dozen books for young readers. Her books have garnered starred reviews, the Eureka! Nonfiction Children's Book Award, the Green Earth Book Award, and the Sigurd F. Olson Nature Writing Award and are on many reading lists. This is her first book for adult readers. Nancy has also authored articles about nature for publications including *Conservationist Magazine*, *Adirondack Life Magazine*, the *Sierra Club Waste Paper*, *Earth Beat*, and the National Resources Defense Council online. She is a member of the Society of Environmental Journalists and is on the council of the International Wildlife Coexistence Network. She lives in New York's Hudson Valley with her husband, Dean, and dog, Boo Radley. Learn more about Nancy Castaldo at https://nancycastaldo.com.

# Index

Abert's squirrels, 11, 38, 66, 175
Abraham, Terry, 93
Ackerman, Diane, 5, 33, 119–120
acorns, 7
Adirondack Mountains, 4–6
air pollution, 169
Alberta, Canada, 37
All Things Squirrel Facebook group, 119
Allen, Jamie, 141–143
Allen, Sophie (dog), 141–142
alligators, 29
ambassadors, 55, 64–65, 73
American red squirrels
    in Adirondacks after loss of chestnuts, 6
    Eurasian red squirrels vs., 70, 79
    food sources for, 7, 17
    hunting data for, 175
    larder hoarding and, 9–10
    in Minneapolis park, 63
    swimming and, 5
American Society for the Prevention of Cruelty to Animals (ASPCA), 60
AMR. *See* antimicrobial resistance
Anatolian ground squirrels, 168
Andrés, José, 87
antimicrobial resistance (AMR), 87
apartment houses for squirrels, 62–63, 140–141
Apennines, 100, 182
Appalachian squirrel stew, 109–110
apples, 65
apps, 42–43
arboreal, 14, 19
Arctic ground squirrels, 152
Arizona gray squirrels, 166
Arnold, Pat, 161
artificial intelligence (AI), 87, 95
Asia Minor ground squirrels, 168
Askew, Henry Garrison, 132

ASPCA. *See* American Society for the Prevention of Cruelty to Animals
Audubon, John J., 17, 132
Ayudándoles a Triunfar, 106–107

bacteria, gut, 84–85, 86–87
Bailey, Vernon, 59
Baldwin, Mary Julia, 70
balsam woolly adelgid, 157
Bangs's mountain squirrels, 168
bark, digestion of, 85
Barry (owl), 125
Bartram, John, 55–56
Bateman, Michael, 133
Bedfordshire, England, 78
Benson, Etienne, 58–59, 70–71, 106, 134, 182
Bergmann's rule, 150
Bertolino, Sandro, 98-103
biased data, 43
BIAZA. *See* British and Irish Association of Zoos and Aquariums
Biden, Joe, 17, 30
Big Squirrel Challenge, 29, 175
biodiversity, 8, 13–14, 18, 21, 23, 24, 42, 47, 91, 106, 134, 136, 143, 151, 167
bird feeders, 115–118, 185
birds, 14, 19, 26, 54, 14, 64, 86–87, 116–120, 124–125, 150
Bisanzio, Donal, 141–142
biters, 65–66
Black Elk (Oglala Sioux holy man), 176
black squirrels, 25, 37, 79
blood chemistry, 50
body size, 69–71, 150–152
Boggs, Joe, 10
Boise, Idaho, 69–70, 117
bomb, 46
Borneo, 12
bounties, 54–55

235

Boutin, Stan, 49
brains, 3
branch tips, cutting of, 10
breeding programs, 91, 160–161
Brickell, John, 53–54
bridges, 180
British and Irish Association of Zoos and Aquariums (BIAZA), 91
Brown, Charlie, 20
Brunswick stew, 109–110
Bryan College, 123
Burroughs, John, 60
burrows, 150, 166–167

caching of food, 2, 7–8, 39, 79, 80–82, 101
Caine, Michael, 85
Calabrian black squirrels, 100–101
California ground squirrels, 149
camera traps, 84, 182
cameras, 61, 91
canopy bridges, 180
Cape ground squirrels, 149–152
car encounters. *See* roadkill
car wires, 121
Carlen, Elizabeth, 42–43
Carolina gray squirrel, 56
Carolina northern flying squirrel, 156–158
Carson, Rachel, 123–124
Carter, Rufus, 87
cartoons, 21, 37–39, 118
Carver, Bill, 118
Celestial Map, 143
censuses, 24–25, 29, 139–141, 143
Center for Biological Diversity, 159
Central Park (New York), 59–64, 139–143, 169, 178
Central Park Squirrel Census, 139–140
Charles, King, 88–89
chestnut trees, 5–6
chewing, 118, 121
chimneys, 121
chipmunks, 17, 21, 115, 119, 133
citizen science, 179, 181
city parks. *See* urban squirrels
Clark, James Rapaport, 135
Clark, William, 131

Cleveland, Grover, 111
climate change
    evolution and, 146, 147–153
    migration and, 133–135, 152
    threatened and endangered populations and, 157, 166
Coates, Peter, 76, 86
coexisting, tips for, 185
cognitive ability, 2–3, 181
college campuses, 70, 122–123, 148–149, 153
Colorado State University, 71
Columbian ground squirrels, 131
comic books, 37–39
communication, 48–49
community science, 42–43, 139–143
Cone, Rhea, 14
conifers, 6, 11
Conservation Cattle Dog Squirrel Squad, 142, 163
contraceptives, 91–92, 95
Copley, John Singleton, 39
CRISPR technology, 93
crossbreeding, 100–101
culling, 87–93, 103, 113, 169
cultural icons, 33–40
*Cumbrian Red* (documentary), 93
*Cyber Squirrel 1* website, 121–122

Darwin, Charles, 147–148, 149
Dean, Nora Thompson, 52–53
deception, 80–82
Decker, Robert Melvin, 5
definition of squirrels, 18–19
Delaware Tribe of Indians, 53
Delgado, Mikel, 9
Delmarva Peninsula fox squirrel, 156
dental research, 146
detection dogs, 142, 163–164
Devitz, Amy-Charlotte, 153
Dewberry, Kerwin S., 160
DiCamillo, Kate, 3
Dickinson, Emily, 35
Dillard, Annie, 33
Dingell, Debbie, 136
diseases, 83, 95, 103, 162
DNA analyses, 146–147

Index    237

DNA modification methods, 93
documentaries, 93
Dodge, Mr. and Mrs. Charles, 62–63
dogs, detection, 142, 163–164
Donaldson, Josh, 28
Downey, Derrick Jr., 63
Downey, Regan, 25–26
drug detection squirrels, 177–178
Dublin, Ireland, 92
Duval, Julian, 132

EarthInvaders.org, 87
earworms, 38
eastern fox squirrels, 9, 11, 17–18, 175
eastern gray squirrels, 9, 17, 25, 30, 42–43, 58–59, 64, 65, 70, 98, 99, 102, 106, 131, 140-142175
eBird, 42–43
ecosystem engineers, 151
ecosystems, 7, 8, 45, 120, 181
Ecuador, 47
Edge, Rosalie Barrow, 116, 182
Emry, Jennifer, 145
Emry, Robert, 145
encephalitis, western equine, 162
endangered species. *See* threatened and endangered species
Endangered Species Act of 1973, 22, 29, 135, 156
Eocene Epoch, 145
Erxleben, Johann C. P., 17
Essex County, New York, 5
Eurasian red squirrels
   before arrival of gray squirrels, 86
   breeding programs for, 91
   college campuses and, 70
   declining numbers of, 83–84
   description of, 79
   in mythology, 35
   nonnative squirrels vs., 79–82, 99–103
   pine species preferred by, 11
   threats to, 49–50
   in United Kingdom, 11
   urban environments and, 181–182
Europe. *See* United Kingdom
evolution, 145–146, 146–153
Eyri (Wales), 95

Facebook groups, 118–119
farming, 65, 146
fat content, 9
Fat Squirrel Week, 71
ferrets, 14
Finland, 45, 181
Finlayson's squirrels, 98, 100–101, 102
Finn, Catherine, 134
fires, 159, 160, 162, 165, 166
Fisher, A.K., 4
five-striped northern palm squirrel, 107
Flaco (Eurasian eagle owl), 125
Fletcher, Nichola, 87–88
Flores, Dan, 57
fluorescence, 49
Flyger, Vagn, 128, 130, 142
flying squirrels, 17, 20–21, 45, 146, 157–158. *See also specific squirrels*
food
   caching of, 2, 7–8, 79, 80–82
   hunting and, 25–29, 60, 87–90, 175
   migration and, 4, 128–135, 149, 152
   omnivory and, 19, 118–120, 149
   squirrels as, 87–88, 109–113
foot size, 145, 150–152
Formosan squirrels, 107
fossils, 145–146
fox squirrels, 47–48, 69–70
Franklin, Benjamin, 55, 57, 77
Franklin Square (Philadelphia), 57–58, 63
Franz, Hilary, 162
Frye Fire (Arizona), 159, 160
fur, fluorescent, 49

Gagnon, Jeff, 11–12
Garfield, James, 111
Genovesi, Piero, 103
Genysys Engine, 87, 95
Gerashchenko, Anton, 94
Germany, 110, 124
germination, 8–9
Gilley, Michelle, 158
Gilliland, Steve, 24, 112
Gladys the Squirrel (mascot), 70
Glendale, Ohio, 37
Glick, Deborah, 27

238    *Index*

globalization, 106
Gmelin, Johann Friedrich, 58
Godman, John, 54–55
Goheen, Jake, 7
golden-mantled squirrels, 66, 82
Goldfarb, Ben, 180
GonaCon, 92, 95
Gooch, Robert, 89
Gottshall, Jonathan, 70
Grand Canyon National Park, 65–66
gray squirrels. *See also specific gray squirrels*
    culling of, 87–89
    delisting of as game species, 175–176
    Eurasian red squirrels vs., 79–82
    as food in North America, 109–113
    as invasive in United Kingdom, 76–79
    in Italy, 98–106
    oral contraceptives for, 91–92
    peak population of in England, 79, 84
    in Philadelphia, 56
    pine martens and, 90–91
    walnuts and, 7
Great Squirrel Migration of 1968, 128–131
Green, Doreen, 38
greeters, 69–74
Grimshaw, John, 91
ground squirrels, 12–14
Guayaquil squirrels, 47
Gunnison's prairie dog, 22
Gunung Palung National Park (Indonesia), 13
gut microbiota, 84–85, 86–87

habitat destruction, 47, 141
Habitat Harmony, 22, 175
habitat loss, 79, 156–157, 158–160, 161–162
habitat requirements, 20
Hadley Bowling Green Inn (Droitwich), 88
Hagen's flying squirrels, 168
Haggett, Gabby, 27
Hale, Dean, 37–38
Hamilton, Rick, 73
Harding, Warren, 55
Hargrove, Christy, 173
Harman, Brian, 28
Harris, Tara, 160–161

Hart, Matthew, 82
Hatt, Robert, 6–7
Haudenosaunee Confederacy, 34
Hawk Mountain Sanctuary (Pennsylvania), 53, 115–116
hazelnut farms, 102
Heaslip, Nancy, 113
heat, dealing with, 149–151
heat dumping, 151
Heinrich, Bernd, 34
Heinrich, Martin, 136
hibernation, 79, 80, 149, 152
Higby, James, 5
HIREC. *See* human-induced rapid environmental change
Hochul, Kathy, 25, 27
Holder, Charles Frederick, 148
housing in Central Park, 62–63, 140
Huckabee, Mike, 110
Hudson Valley (New York), 111, 121, 128–131, 178
Huettmann, Falk, 30
human-induced rapid environmental change (HIREC), 134, 147
hunting, 25–29, 60, 87–90, 175
hydration, 50

immigrants, 56, 79, 113
importation, prohibition of, 77
iNaturalist, 42–43
India, 36
indicator species, 153, 162
Indonesia, 13
Inhofe, James, 24
insulation, 118
intelligence, 80–81
International Colloquiums on Squirrels, 24, 44
invasive species
    conservation of native species and, 90–92, 94–96
    culling of, 87–93, 107
    documentary about, 93–94
    history of introductions and, 76–79, 98–100
    insects, 157, 160

in Italy, 98–106
in Japan, 107
native species vs., 79–83, 87
squirrel health and, 83–87, 157
in United Arab Emirates, 107
in United Kingdom, 11
invasivorism, 87
island populations, 29–30, 45, 142, 162
Isle of Wight, 29
Italy, 11, 34, 39, 49–50, 79, 96, 98–106

Jack (dog), 163–164
Jacobs, Lucia, 9
Japan, 19, 107, 156, 181
Japanese giant flying squirrels, 19
Japanese pygmy dwarf squirrels, 19
Japanese squirrels, 38, 45
Jerolmack, Colin, 143
Johnson, Jeremy, 53
Jones, Terry L., 112
*The Joy of Cooking* (Rombauer), 110–111
Julia Davis Park (Boise), 69–70

Kaibab squirrels, 66
Kalahari grassland, 149–152
Kelly, John, 35, 174
keystone species, 2, 7–8, 10–11, 12, 21–22, 174
Kingston, Danny (Food Urchin), 88
Koprowski, John, 152–153, 159, 181
kuks, 48

La Ragione, Roberto, 84–85
Lacey Act of 1900, 29, 77
Lake George )New York), 4–5
larder hoarding, 7, 9–10
Large Binocular Telescope (Arizona), 159–160
Lawrence, Jennifer, 111
Lawrence, D.H., 98
lawsuits, 159, 162
Lawton, Colin, 44
Lee, Hannah, 149
Lenape people, 52–53
Leporipoxviruses, 83
Lewis, Meriwether, 131
Linden, Autumn, 73

Lombard Nature Reserve (South Africa), 149–152
London, England, 78, 169
Longfellow, Marian, 63
longleaf pine, 11
Loring Park (Minneapolis), 63
lymphoma, 49–50

Malaysia, 12
mange, 162
maple sugaring, 34
Marie-Victorin, Frère, 34
Marvel comic books, 37–38
Mary Baldwin University (Virginia), 70
Maryville College, 148–149
mascots, 37, 45, 70, 178
mast years, 49, 129–130, 131, 133, 134
Matthews, James, 109–110
Mazzamuto, Maria Vittoria, 104–106
McCartney, Paul, 85–86
McCleery, Robert, 47–48, 121
McKeown, Christy, 173
McMaster, Doug, 89
McRae, Thaddeus, 23, 48–49
melanism, 43
Menchetti, Mattia, 181
Mercer, John M., 146–147
Mersea Island (England), 30
Michigan, University of, 153
microbiota, gut, 84–85, 86–87
midden-gifting, 10, 79
middens, 7, 10, 18, 79
migration
    in 1833, 132
    in 1968, 128–131
    climate change and, 133–135, 149, 152
    for nuts, 4
    theories on causes of, 4, 129–130, 132–133
military terms, 39
Mills, Heather, 85
Minneapolis, Minnesota, 63
*Miopetaurista* genus, 146
Miyazaki, Hayao, 38
moans, 48
modeling, 12–13
Moeller, Anna, 26

240  *Index*

Mohave ground squirrels, 167
mongoose, 76
Mori, Emiliano, 181
Morris, Alice, 167
Mount Graham red squirrel, 158–161
Mungo (squirrel), 77
Munroe, Karen, 47, 96, 155–156
Murray, Joan, 35
muskrats, 29
mutualisms, 10–11
myths, 34-36

Nasdaq disruptions, 122
National Animal Rights Association (NARA), 92
National Botanic Gardens (Dublin), 92
national parks, 64–65
National Squirrel Appreciation Day, 172
National Wildlife Institute, 103
National Wildlife Research Center, 92
Native Americans, 34, 52–53, 109, 176
Needwood Forest (Staffordshire), 78
*Nelaballi, Swapna*, 13
New England Flying Squirrel Network, 179
New York Sun, New York, 17–18, 59–64, 139–143, 178
*The New York Sun* (newspaper), 60–61
Newagen Seaside Inn (Maine), 33, 123, 125
Niko (squirrel), 106–107
Nishida, Atsuko, 38
Norse mythology, 35–36
northern flying squirrels, 21, 156–158
northern Idaho ground squirrels, 167
nuisance species, 28, 165
Nutkin, Squirrel (Potter character), 82–83

Ogwen Valley (Wales), 83
Oliver, Mary, 35, 115
Olson, Vickie, 65
omnivory, 19, 118–120, 149
owls, 124–125
oxalate-degrading bacteria, 85

Pallas's squirrels, 98, 102–106
palm squirrel, five-striped northern, 107
Paris, France, 77–78

parks. *See* national parks; urban squirrels
passenger pigeons, 29
Patchwork Traditional Food Company, 87
Pauley, Sara Parker, 136
Penn, William, 53
Penrith & District Red Squirrel Group, 93
perishability, 9
Perlut, Noah, 71–73, 116
Persian squirrels, 21
pests, squirrels as, 54–55
*Petaurista* genus, 146
Peters, William David, 9
pets, 55, 106, 177
Philadelphia, Pennsylvania, 52–59, 63
Phoenix Zoo, 160–161
phone trees, 141
photography, 174
Piedmont, Alabama, 122
Piedmont, Italy, 99- 102
Pikachu, 38
Pincheira-Donoso, Daniel, 134
pine martens, 90–91
pine species, preferences for, 11
Pisgah Penny (squirrel), 172
Pittman-Robertson Wildlife Restoration Act, 136
Platt, John, 103
Pliny the Elder, 34, 39
pocket gophers, 20
poetry, 35
poisonings, 123–125
pollution, 169, 170, 180
ponderosa pine, 11
poplar plantations, 102
population numbers. *See also* censuses
    Central Park and, 139–143
    from hunters, 175
    need for more, 179–180
    statistics on, 24–25, 29, 134–135
Potter, Beatrix, 82–83
power lines, 121–123, 160
pox, 83, 95
prairie dogs, 14, 20–22, 92, 155
predator satiation, 129
predator-prey relationships, 119–120
predators, 19, 90–91, 118–120, 149

prey, 13–14, 19, 21, 23, 118–120, 149
problem-solving abilities, 3
Project Squirrel, 71–73
Purdue University, 70

Qian Xuan, 39
quaas, 48

Race Street (Philadelphia), 57
Ratatoskr, 35–36
Rathore, Kalyan, 36
RAWA. *See* Recovering America's Wildlife Act
Rawlings, Marjorie Kinnan, 109–110
Recovering America's Wildlife Act (RAWA), 135–137
Red Haven, Project, 87
Red Squirrel Appreciation Day, 89, 172
Red Squirrel Studbook network, 91
Red Squirrel Survival Trust, 88–89, 91–92
Red Squirrel Week, 85–86
red squirrels, 158–161. *See also* American red squirrels; Eurasian red squirrels
Red Squirrels Northern England, 92
Regent's Park (London), 78
rehabilitators, 173–174, 177
reintroduction efforts, 30, 83
renewable resources, animals as, 28
Renn, Emily, 22, 175
Reribu. *See* tufted ground squirrels
Rewilding Ukraine, 46
*Rhizosphaera* needle cast, 18
Rich, Terry, 117
Richardson's ground squirrels, 152
Richmond Park (Surrey), 78
roadkill, 128–129, 133, 180
Robertson, Wade, 111–112
rock squirrels, 65–66
Rodentia, 9, 18, 121
rodenticides, 124–125
Rogue Detection Teams, 163
Rombauer, Irma, 110–111
*Romeo and Juliet* (Shakespeare), 82
Rose, Adam, 101
Roth, V. Louise, 146–147
Rusin, Mikhail, 46

Russell, Herbrand, 78

Sage, Russell, 62–63
Saint Helens, Mount (Washington), 19–20
Salten, Felix, 83
Salzman, Jason, 113
Sassafras Street (Philadelphia), 57
Save Our Squirrels, 87
scatter hoarding, 7, 13, 81
scatterbrained, 40
*Sciuridae*, 9, 18, 19
Scribble a Squirrel, 85–86
sculptures, 36–37
SDM. *See* species distribution models
Secret Squirrels, 39
seed dispersal, 2, 6–7, 10–12
seed predators, 10–13
sentinel species, 49–50
Sequin Squirrel Trail (Texas), 37
Seton, Ernest Thompson, 2, 3–5, 14, 59, 155, 179
Shannon, Jim, 94
Sheehy, Emma, 90
Shull, Alisa M., 158
Siberian chipmunks, 21, 102
Siberian flying squirrels, 45
Signorile, Lisa, 84, 102
Silo (London), 89
Silver, Robin, 159
Silverman, Sarah, 2
Simmons, Jessica, 14
Sironi, Francesca, 101
size, defining squirrels and, 19
Skabeyeva, Olga, 94
Skeeter the Gray Squirrel, 73
skeletons, 145–146
Skelwith Fold Caravan Park, 90
Smallwood, Peter D., 9
Smiley, Daniel, 130
Smith, John, 62
Smithsonian Institute, 130, 145, 146
Snell-Rood, Emilie, 153
sniffer squirrels, 177–178
Snyder, Kristina, 176
social media, 37–39, 60, 82, 106, 116
southern flying squirrels, 21, 182

Sparky the Squirrel, 122
spatial chunking, 3
species distribution models (SDM), 12–13
Speed, Rob, 123
sperm, 152
Spieler, Marlena, 87–88
splooting, 151
spotted squirrels, 66
Spragins, Pam, 173–174
spray-painting of squirrels, 178–179
spruce trees, 17–18, 22–23
Squirrel Agent: Project Red Haven, 95
Squirrel Appreciation Day, 180
squirrel brains, 3, 40
Squirrel Census Phone Tree, 140–141
squirrel feeders, 125–126
Squirrel Girl (Marvel character), 37–38
Squirrel Haiku Contest, 35
Squirrel Haters of America Facebook group, 116, 118
Squirrel of the Week features, 70
Squirrel Scramble, 25
Squirrel Sighters, 42, 140
Squirrel Week, 174
squirrelpox virus (SQPV), 83, 95, 103
squirrel-rehab.org, 173–174
Squirrels are Awesome Facebook group, 116, 119
*Squirrels of the World* (Thorington et al), 9
*Squirrels R Us* website, 173
squirrely, origins of word, 40
Staffordshire, England, 78
statistics, 137, 155. *See also* censuses; population numbers
Steele, Michael, 3, 9
Steiner, Moriz, 30
sterilization, 106
Stieglitz, Alfred, 5
Stihler, Craig, 157
Stith, Ned, 110
Stokes, Mike, 153
Stupinigi, Parco di (Nichelino), 99, 102
subspecies, discovery of new, 181–182
Surrey, England, 78
survivorship statistics, 23
Swayser, Renae, 20–21

swimming, 4–5, 96

tails, 34–35, 150
*Tamiasciurus* genus, 17
tannins, 9
territoriality, 9–10, 17, 65, 79
terrorists, squirrels as, 94, 103, 119
Thomas, Cris, 121–122
Thomas Darling Preserve (Pennsylvania), 21
Thoreau, Henry David, 8, 173
Thorington, Kelly, 174
Thorington, Richard, 9, 145, 174
threatened and endangered species
    Arizona gray squirrels as, 166
    Carolina northern flying squirrel and, 156–158
    criteria for listing as, 29
    delisting of, 175–176
    Delmarva Peninsula fox squirrel and, 156
    European squirrels on list of, 46
    Mohave ground squirrels as, 167
    Mount Graham red squirrel and, 158–161
    northern flying squirrels as, 20–21
    northern Idaho ground squirrels as, 167
    outside of Untied States, 168
    protection of, 135–137
    Washington ground squirrels as, 163–166
    western gray squirrels as, 161–163
Timmy Tiptoes (Potter character), 76, 83
Tom Girl (squirrel), 72
torpor, 79
Totoro, 38
trapping, 72, 92, 105, 107, 178
tree damage, 85
tricks, 64
Truman, Harry, 55
Trump, Donald, 137
Tucker, Tommy, 64
tufted ground squirrels, 12, 13
tunnels, 14
Tyning, Tom, 179–180

Ukraine, 46, 94
ultrasonic noises, 158
*The Unbeatable Squirrel Girl*, 37, 38

United Arab Emirates, 107
United Kingdom. *See also specific countries*
   competition between squirrels in, 79–84
   culling gray squirrels in, 87–90
   declining numbers of red squirrels in, 83–84
   gray squirrel contraceptives and, 91–92
   importation of gray squirrels to, 76–79
   invasive species in, 11
   pine martens and, 90–91
   squirrel diseases and, 83, 95, 103
   supporting red squirrels in, 85–87, 93, 94–96
University College (London), 84
University of New England, 71–73
urban squirrels
   air pollution and, 169–170, 181
   in Boise, 69–70
   history of in United States, 52–53
   in Minneapolis, 63
   in New York City, 59–64, 139–143, 178
   in Philadelphia, 57–58, 63
Utica University, 122

vaccines, 92, 95, 107
Vanek, John, 42–43
Veigel, Steven, 132–133
video games, 180
Villa Gropallo, Parco di (Genoa Nervi), 99
vocalizations, 48–49

Wallis, Kelly, 165–166

walnuts, 3, 7, 126, 131
Warrington, Miyako, 150–152
wars, 46–47
Washington ground squirrels, 163–166
Waterman, Jane, 150–152
Watson, Winslow, 5
western gray squirrels, 135, 161–163
White River Formation, 145
White Squirrel Day, 172
white-bellied northern flying squirrels, 21
Wiggins, Elise, 109, 112–113
Wild Harvest Initiative database, 175
Wild Meat Company, 89
wildfires, 135, 155, 159–162, 164, 165, 166, 167
Willcox, William R., 62
Wirth, Theodore, 63
Woburn Abbey (Bedfordshire), 77–78
Wong, Catherine J., 180
woodchucks, 17
World Champion Squirrel Cook Off, 175
Wu, Tony, 45

Yee, Collette, 163–165
Yeison (Venezuelan migrant), 106–107
Yellowstone National Park, 65
Yolen, Jane, 80
Yukon Territory, Canada, 49

Zimmern, Andrew, 111
Zion National Park, 65

GPSR Authorized Representative: Easy Access System Europe - Mustamäe tee 50, 10621 Tallinn, Estonia, gpsr.requests@easproject.com

www.ingramcontent.com/pod-product-compliance
Lightning Source LLC
Chambersburg PA
CBHW072337260426
43938CB00034B/476